WEYERHAEUSER ENVIRONMENTAL BOOKS

William Cronon, Editor

Weyerhaeuser Environmental Books explore human relationships with natural environments in all their variety and complexity. They seek to cast new light on the ways that natural systems affect human communities, the ways that people affect the environments of which they are a part, and the ways that different cultural conceptions of nature profoundly shape our sense of the world around us.

WEYERHAEUSER ENVIRONMENTAL BOOKS

The Natural History of Puget Sound Country
by Arthur R. Kruckeberg

*Forest Dreams, Forest Nightmares: The Paradox of Old Growth
in the Inland West* by Nancy Langston

Landscapes of Promise: The Oregon Story, 1800–1940
by William G. Robbins

*The Dawn of Conservation Diplomacy: U.S.-Canadian Wildlife Protection
Treaties in the Progressive Era* by Kurkpatrick Dorsey

*Irrigated Eden: The Making of an Agricultural Landscape
in the American West* by Mark Fiege

*Making Salmon: An Environmental History of the Northwest
Fisheries Crisis* by Joseph E. Taylor III

George Perkins Marsh, Prophet of Conservation
by David Lowenthal

*Driven Wild: How the Fight against Automobiles Launched the
Modern Wilderness Movement* by Paul S. Sutter

The Rhine: An Eco-Biography, 1815–2000
by Mark Cioc

Where Land and Water Meet: A Western Landscape Transformed
by Nancy Langston

*The Nature of Gold: An Environmental History of the
Alaska/Yukon Gold Rush* by Kathryn Morse

Faith in Nature: Environmentalism as Religious Quest
by Thomas Dunlap

WEYERHAEUSER ENVIRONMENTAL CLASSICS

The Great Columbia Plain: A Historical Geography, 1805–1910
by D. W. Meinig

Mountain Gloom and Mountain Glory:
The Development of the Aesthetics of the Infinite
by Marjorie Hope Nicolson

Tutira: The Story of a New Zealand Sheep Station
by Herbert Guthrie-Smith

A Symbol of Wilderness: Echo Park
and the American Conservation Movement
by Mark W. T. Harvey

Man and Nature: Or, Physical Geography as Modified by Human Action
by George Perkins Marsh; edited by David Lowenthal

Conservation in the Progressive Era: Classic Texts
edited by David Stradling

CYCLE OF FIRE BY STEPHEN J. PYNE

Fire: A Brief History

World Fire: The Culture of Fire on Earth

Vestal Fire: An Environmental History, Told through Fire,
of Europe and Europe's Encounter with the World

Fire in America: A Cultural History of Wildland and Rural Fire

Burning Bush: A Fire History of Australia

The Ice: A Journey to Antarctica

FAITH IN NATURE

Environmentalism as Religious Quest

THOMAS R. DUNLAP

Foreword by William Cronon

UNIVERSITY OF WASHINGTON PRESS

Seattle and London

Faith in Nature: Environmentalism as Religious Quest
has been published with the assistance of a grant from the
Weyerhaeuser Environmental Books Endowment, established
by the Weyerhaeuser Company Foundation, members
of the Weyerhaeuser family, and Janet and Jack Creighton.

Copyright © 2004 by University of Washington Press
Designed by Pamela Canell
Printed in the United States of America
10 09 08 07 06 05 04   5 4 3 2 1

University of Washington Press
PO Box 50096, Seattle, WA 98145, U.S.A.
www.washington.edu/uwpress

Library of Congress Cataloging-in-Publication Data
available from the Library of Congress
ISBN 0-295-98397-3

The paper used in this publication is acid-free and recycled from 10 percent post-
consumer and at least 50 percent pre-consumer waste. It meets the minimum
requirements of American National Standard for Information Sciences—
Permanence of Paper for Printed Library Materials, ANSI z39.48–1984. ⊗ ☺

DEDICATION

THIS ONE IS FOR MY FRIENDS, the whole kit and caboodle, for all they have done and for all the good times. The line starts with schoolmates and college classmates, goes on to the boys in the barracks, and reaches a high point with the friends I found in becoming and being an historian. Some deserve special mention. The crew from high school included Larry, a stand-up guy; Barbara, whose conversations and cutting remarks helped me grow up; Sandy, a solid person and still our class organizer; and Bunny (a.k.a. Anne), who was always willing to argue. In graduate school I met Pete Cannon, and we shared an apartment and managed the little complex we lived in for two years and spent our summers painting houses and apartments to supplement our G.I. Bill benefits. He and his wife, Marsha, and my wife, Susan, and I have been friends for almost thirty years now, and I can only hope for many more laughs and more great vacations. Harold Livesay and his wife, Cyndy Bouton, have been our friends for more than twenty years and my colleagues for almost that long. They watched our daughter grow to adulthood as we watched their son, and shared Thanksgivings, other meals, and all the gripes and jokes and good times that go with life in this business. I owe them, particularly Harold, more than I can reasonably put in a dedication. Then there is Bill Cronon, a model professional colleague and friend. I owe him much for this book but more for friendship.

Friendship is a sheltering tree.

CONTENTS

FOREWORD

Searching for the God in All Things

William Cronon

IS ENVIRONMENTALISM A RELIGION? Even to ask the question may seem outlandish. Those who follow more traditional faiths may think it nearly heretical to identify religious tendencies in an intellectual and political movement that is often so materialist and anti-metaphysical in its impulses. To worship the earth and its creatures instead of a transcendent deity might seem the very antithesis of religion. Looked at from this perspective, environmentalism might more appropriately be labeled a form of atheism, and surely many of its followers would be quite content with that label. Indeed, at least a few environmentalists will undoubtedly be disturbed by the suggestion that concern about pollution or biodiversity or global climate change might be anything other than purely rational. What could be more practical or worldly than caring about the survival of our earthly home? Why would religion have anything to do with that?

And yet environmentalism does share certain common characteristics with the human belief systems and institutions that we typically label with the word *religion.* It offers a complex series of moral imperatives for ethical action, and judges human conduct accordingly. The source of these imperatives may not appear quite so metaphysical as in other religious traditions, but it in fact derives from the whole of creation as the font not just of ethical direction but of spiritual insight. The revelation of seeing human life and the universe whole, in their full interconnected complexity, can evoke powerful passions and convictions ranging from the mystical to the missionary. Certain landscapes—usually the wildest and most natural ones—are celebrated as sacred, and the emotions they inspire are akin

xi

to those we associate with the godhead in other faith traditions. Much environmental writing is openly prophetic, offering predictions of future disaster as a platform for critiquing the moral failings of our lives in the present. Leave out the element of divine inspiration, and the rhetorical parallels to biblical prophecy in the Hebrew and Christian traditions are often quite striking. Maybe most important, environmentalism is unusual among political movements in offering practical moral guidance about virtually every aspect of daily life, so that followers are often drawn into a realm of mindfulness and meditative attentiveness that at least potentially touches every personal choice and action. Environmentalism, in short, grapples with ultimate questions at every scale of human existence, from the cosmic to the quotidian, from the apocalyptic to the mundane. More than most other human endeavors, this is precisely what religions aspire to do.

Thomas Dunlap's *Faith in Nature: Environmentalism as Religious Quest* is an extended meditation on these convergences between environmentalism and other religious traditions. Because some potential readers of this book may initially be surprised by—or even resistant to—its core claim that our understanding of environmentalism can be substantially enriched by exploring its religious aspects, it may be even more helpful than usual to know something about why this particular author was drawn to this particular project in the first place. Readers should know that Tom Dunlap is among the leading environmental historians and historians of science in the United States. He has made substantial contributions to our understanding of environmental thought and politics, starting with his pioneering book *DDT: Scientists, Citizens, and Public Policy* and ranging outward from there to such disparate topics as the history of American predator control and the comparative history of nature writing in the United States, Canada, Australia, and New Zealand. Originally trained as a chemist, he has a rigorous understanding of science and appreciates its vital importance to environmental thought. He writes as someone who cares enormously about environmental issues, and who has the technical skills to explore them in their full complexity.

But Tom Dunlap is also a devout practicing Catholic who believes that the insights of religious revelation need not be necessarily at odds with the insights of scientific investigation (including Darwinian evolution). As he explains in the opening section of *Faith in Nature,* he grew up in the White Mountains of New Hampshire as the child of parents who taught him science and religion in equal measure. A passionate birdwatcher from an early age, he spent his boyhood hiking trails that carried him ever deeper into the wonders of a natural world that has been every bit as much a source of ultimate meaning in his life as it has been for so many others of us who call ourselves environmentalists.

Because he was initially drawn to a scientific career before eventually making his way to the history of science and environmental history, he found himself pursuing a long personal quest to discover how his religious faith and his science might inform each other. Although some might find the two irreconcilable and conclude that one or the other must be rejected outright, that is not Tom Dunlap's way. Among his most engaging and attractive qualities is his perennial openness to new ideas and new questions, many of which seem to arise most fruitfully from the apparent contradictions that initially seem most troubling. Already ecumenical by inclination, his appreciation for religious complexity has undoubtedly been enriched still further by his long, happy marriage to his wife, Susan Miller, who is as committed to her Judaism as he is to his Catholicism. Tom and Susan have chosen to raise their daughter, Margaret, in the Jewish tradition even as Tom himself has continued to attend Mass—all the while studying science and Darwin and writing about the history of environmentalism.

This is, in short, a complicated human being who takes nothing about religion or human values for granted. I've known Tom Dunlap for nearly twenty years and have been aware of his personal intellectual journey for a long time. I still remember the hot summer afternoon when it first occurred to both of us that a short book on environmentalism and religion might be something he'd be uniquely qualified to write. We were standing beside the Wisconsin River just north of Aldo Leopold's famous Shack and talking about the controversy that had recently erupted concerning an essay I had written about wilderness. One of the premises of that essay was the familiar academic observation that eighteenth- and nineteenth-century romanticism redefined wild nature so as to make it a repository for religious values that had formerly been more associated with cathedrals and sacred texts. By this argument, environmentalism is a direct descendant of romanticism's success in mapping secularized religious values onto the natural world.

Because Tom understands the implications of this argument better than most, our conversation wandered from wilderness to romanticism to religion along a path that led finally back to Aldo Leopold himself, who wrote as eloquently about human ethical obligations and spiritual debts to the natural world as any author in the environmental canon. The more we talked, the more it seemed to me that Tom's own religious journey had given him a wealth of insights into environmentalism that deserved systematic treatment in a book. I suggested the idea, and it was instantly clear that it struck a chord for him. Not long afterward, he put together a proposal and a table of contents, and we soon had a contract. For the past several years, few months have passed without extended communications between the two of us, whether by phone or e-mail or in person, about the progress of the project. It has been among the most satisfying editing experiences I've ever had, and *Faith in Nature* is its end product.

I share all this personal information simply to underscore what should be obvious to anyone who gives this book a fair-minded reading. *Faith in Nature* is anything but a muckraking attack on environmentalism or an effort somehow to diminish the importance of environmental values by ascribing them to religious conviction. Quite the contrary. For Tom Dunlap, religions are among the most important places where human beings explore the deepest meanings of their lives and the ultimate mysteries of the universe. To describe environmentalism as an emerging religious tradition is thus much more an act of praise than of criticism . . . though readers will find plenty of criticism here as well, mainly pointing toward the difficulties that can arise when environmentalists fail to grapple with key challenges that all religious traditions must eventually confront.

Speaking for myself—though I'm pretty sure Tom would not disagree with this—I would go further still. I would argue that environmentalism and its various antecedents represent one of the most sustained and creative efforts over the past two centuries to translate core religious values so as to demonstrate their continuing relevance to a modern age that often seems relentlessly secular, materialist, and irreligious. Among the most brilliant achievements of this environmental tradition has been its success in embracing a materialist vision consistent with modern science at the same time that it finds deep spiritual meaning at the heart of a material universe that might otherwise seem soulless. In its holism, its communitarianism, its vision of the sacred, and in the claims of individual and collective moral responsibility that it derives from the intersection of these other values, environmentalism goes a long way toward imagining what a "natural religion" might look like in the modern world. *Faith in Nature* offers a generous and thought-provoking sketch of how this environmental religious tradition has emerged over time, and where it might be headed in the future.

ACKNOWLEDGMENTS

BILL CRONON GUIDED THIS BOOK from its beginnings in a letter I wrote him after hearing the session on "The Trouble with Wilderness" at a meeting of the American Society for Environmental History. After a few letters he suggested I write a book about "environmentalism as a religion." E-mails, drafts, and letters began to pile up, supplemented with some intense conversations. Bill provided continual thoughtful criticism and, since my involvement went beyond the professional and "historical," many comments about my own perspective and how my growing up was involved in my views. On this project I owe him as friend as much as I owe him as professional colleague and editor; it was the thought of our friendship through this project that suggested the dedication.

Despite close collaboration, though, I take full responsibility for all errors, oddities, and strange choices. Julidta Tarver of the University of Washington Press was helpful, patient, and all those other adjectives that authors apply to editors who make their lives easy by friendship and hard work. Michael Cohen set an example of academic generosity by volunteering to read a version, doing it quickly, and offering useful and trenchant comments, some of which I incorporated, others I filed away for future use.

My wife, Susan Miller, and our daughter, Margaret Miller Dunlap, were their usual wonderful selves. On a more professional level, I relied, as every academic must, on my university. The Texas A&M library, particularly the interlibrary loan office, provided essential support, and my department heads, Julia Blackwelder and then Walter Buenger, unfailing help.

FAITH IN NATURE

Preferring a search for objective reality over revelation
is another way of satisfying religious hunger.

E.O. WILSON

Talk of mysteries! Think of our life in nature—daily to be shown
matter, to come into contact with it—rocks, trees, wind on our
cheeks! the *solid* earth! the *actual* world! the *common* sense!
Contact! Contact! Who are we? *Where* are we?

HENRY DAVID THOREAU

The intangible values of wilderness are what really matter,
the opportunity of knowing again what simplicity really means,
the importance of the natural and the sense of oneness with the
earth that inevitably comes within it. These are spiritual values.
They, in the last analysis, are the reasons for its preservation.
Wilderness will play its greatest role, offering this age a familiar
base for exploration of the soul and the universe itself.

SIGURD OLSON

Save the Planet!

The habit of bluntly opposing nature and culture has only
gotten us into trouble, and we won't work ourselves free
of this trouble until we have developed a more complicated
and supple sense of how we fit into nature.

MICHAEL POLLAN

There is the question of "whether the philosophy of industrial
culture is not, in its ultimate development, irreconcilable
with ecological conservation. I think it is."

ALDO LEOPOLD

Introduction

> At bottom, the whole concern of religion is with the manner
> of our acceptance of the universe.—William James

It was the manner in which some historians reacted to William Cronon's "The Trouble with Wilderness" that first got me thinking, consciously at any rate, about the topic that turned into this book. Cronon had made an academic point: because of certain historical processes and events we give the name "wilderness" to areas of land with certain characteristics, and by doing that shape the way we look at the world and what we value in it. Members of the audience had argued against Cronon's point and his view of wilderness, and with more passion than academics usually display unless their own work is under attack. Wilderness, they insisted, was something real, not a name we give to some parts of the world, and by seeing "wilderness" as an historical construct, they argued, Cronon gave anti-environmentalists ammunition to argue against saving wilderness. This was aid and comfort to the forces of evil—or Evil. A comparison came to mind. The historians seemed like Christian fundamentalists listening to an historical-critical talk on the Bible. The speaker might be a Christian by his own lights, but in viewing the Holy Scriptures as products of the times and the recorders of God's Words, he showed himself an apostate. So it was with Cronon putting our ideas of wilderness in historical perspective. Impressed by the environmentalists' passion, I looked more closely at its source. I already knew it ran deep. Even if I were not an environmental historian I would have seen the news

3

stories about people lying down in front of bulldozers to save ancient forests, living in the tops of trees to keep loggers from redwoods, and running small boats in front of whaling ships to block the harpooning of whales. As an environmental historian, I also knew that many middle-aged, middle-class, respectable citizens not only vote for candidates who promise to protect the environment, but give money to the cause, sign petitions, and applaud—even if they do not join in—radical actions like tree-spiking.

Environmental degradation certainly accounts for a lot of that passion. Only the terminally optimistic believe environmental problems will just go away. Everyone else looks at denuded mountainsides and eroded fields; notices the air and the water have a lot more "body" than God gave them; reads about rising rates of extinction, a decaying ozone layer, and a changing climate; and wonders how much trouble we really are in. But the dangers to humans do not account for all of the environmentalists' passions and actions. The environmental movement demands more than natural beauty, preserving human health, or even building a sustainable society—demands more, even, than reform. Reform movements seek new laws that will bring justice, and they tell you how to vote, whom to boycott, and how to work for the cause. They appeal to moral standards but are fundamentally concerned with a more just society. Environmentalism campaigns for new laws, but it also gives moral weight to the apparently trivial decisions of daily life. It tells you what kind of grass to put in the front yard, how to get to work, even what kind of diapers to put on the baby. It makes a brick in the toilet tank an expression of virtue. It asks not just that we change policies or even our habits, but that we change our hearts, not just that we recycle papers, cans, and bottles, but that we form a new relationship with nature. Finally, it invokes the sacred, holding some areas and species in awe and finding in wilderness the opening to ultimate reality.

Turning over books and arguments and ideas, I began to think about environmentalism as an expression of the human impulse toward religion, which, admittedly, sounds odd. Environmentalists are enthusiastic, but there is no First Church of Darwin, Environmentalist, and the Sierra Club does not promise heaven or enlightenment (though the calendar with the subscription has devotional pictures). Environmentalists do not, as environmentalists, recite creeds, hold services, or call on mystical forces. At least they do not call their positions "creeds" or their acts "services," and most shrink from the idea of mystical disciplines, except, possibly, for individual self-improvement. They are solid, respectable, intelligent citizens, holding to science and reason.

Consider, though, William James's definition of religion in *The Varieties of Religious Experience* as the "belief that there is an unseen order, and that our supreme good lies in harmoniously adjusting ourselves thereto." Religion arises from the common human "sense that there is *something wrong about us* as we naturally stand. The solution is a sense that *we are saved from the wrongness* by making proper connection with the higher powers."[1] That perspective makes the religious impulse part of the human condition and religion our framework for answering ultimate questions: why are we here, and what must we do to fulfill our purpose in life? It does not (necessarily) involve God (or gods) or devils, an afterlife, revelations from On High, prophets, or miracles. A religion need not be coherent, provide a full set of answers to life's mysteries, or offer a path to salvation. Religion is how we make (ultimate) sense of our lives in the context of the universe.

Applying a religious perspective to environmentalism may make nonbelievers and believers uncomfortable. Those who see religion as superstition may fear this point of view is a tactic to discredit the movement by associating it with wooly minded mysticism. Those who hold strongly to one of the conventionally defined faiths may feel it reduces their beliefs to a psychological impulse. Certainly there have been newspaper opinion columns with headings like "Green worshippers can keep their religion" and letters describing environmentalism as a "a new religion, a new paganism that worships trees and sacrifices people." On the other hand, many in the modern world scorn religion as a crutch of those too weak to face reality.[2] This book takes neither of those positions. It examines environmentalism, its roots in the culture, and its development as a movement in religious terms—as a way of accepting the universe—in order to think about its foundations and the source and depth of its passions.

That said, there are still reasons to be uncomfortable. It is one thing to read an anthropological study on the worldview of the Longlost tribe that lives on the Upper Boondocks River in Darndrearyland, quite another to find that sort of analysis trained on your own beliefs. This study is not anthropological, but it does raise unsettling thoughts: perhaps, even in our secular, scientific, rational age, even among those who cleave to science and hold rational argument dear, people ultimately rely on something beyond rational demonstration. Perhaps we, like the benighted people sacrificing goats in the backyard or dressing up for church on Sunday, build our worldview not on certain knowledge but on faith—if only faith in science and our own reasoning powers. On the other hand, if environmentalism strikes to the level of religion, if it speaks to how we should live, then perhaps it has something important to say to conventional creeds.

This analysis cuts close to the bone in another way. The things that meet and mingle here—"religion," "science," and "nature"—meet and mingle in each of our lives. Some of us receive as children a formal creed we hold for life or spend a lifetime getting away from. Even those raised without a declared faith deal with ultimate questions and confront conventional believers. We all learn the point of view and other cultural baggage associated with science, even if we never study its theories, and we live in a society that uses science to decide what is "out there" and what constitutes knowledge about what it is. We all encounter nature, which overwhelms some of us, interests others, and repels a few. Necessarily, we work out apparent and real clashes between religion, science, and nature in our own lives, fitting together what we are taught, what we understand, and what we feel. Some of us find science, family belief, and interest running together; others battle contrary currents. A child fascinated with nature but raised in a religion that rejects this world, or who encounters science as destructive analysis, will have some trouble making sense of things.

My own case suggests the usual complexities. I encountered religion, science, and nature as, respectively, my mother's creed (in which I was raised), my father's profession and enthusiasm (ditto), and the land beyond the backyard (which I found fascinating). My mother's Irish Catholicism, complete with rosaries for presents and after-school catechism classes for penance, provided not just doctrine, but a worldview and sensibility that colored my approach to almost everything. (Those who think this too strong should ask any former or recovering Catholic. My friend Cyndy, who will walk across the street to avoid a Catholic church, admits, with a sigh, that after thirty years she still has the Mark of the Beast on her forehead.) I did not, though, grow up in a Catholic society or even a fully Catholic family. My father got pork chops on Friday while the rest of us dug into tuna casserole, and, though he never went to church or mentioned his beliefs, he carried other marks of his descent from (as he once put it) "a long line of bigoted Ulster Protestants" (he had hastily added that this was a redundant description). Besides, we lived in rural, Protestant America, where five verses from the King James Bible and class recitation of the Lord's Prayer (the Catholics mumbling where the words were different) graced each day in grade school. Here a local carpenter could declare his satisfaction with a job by saying, "That's square and Protestant."

Because science lacked a baptismal ceremony, it came into my life after Catholicism, but Dad did his best to make up the lost ground. He talked science at the dinner table, brought me books and gave me a microscope for Christmas, and encouraged me to become a chemist—preferably a paper chemist. He started taking me to the lab when I was about six and into paper mills when I was in junior high. School reinforced my father's enthusiasm for science with natural history lessons and pamphlets and then courses organized by discipline. Entering high school as *Sputnik* went up, I found my country's urgings added to the school's. Being a dutiful child, I majored in chemistry in college and did a couple of graduate years in high-temperature thermodynamics before I decided I did not want to spend my life in a lab. Even when I found my way to history, science dogged my steps. My advisor suggested that with my background, I should write a history of DDT for my master's thesis, and I have worked on the topic of science and people's ideas about nature ever since.

Nature entered my life quite early, for my first memories include exploring what seemed to a three-year-old a wilderness of grass beyond the yard and past the end of the block, fields that would soon succumb to the expanding post-war Chicago suburbs. We left for New Jersey before the march of progress turned fields to lawns, and I spent my grade school years outside a small potato-farming village, surrounded by fields and woods, with a stream just down the road. I caught frogs and turtles; found snakes; saw deer and rabbits; started fishing with poles, string, and pins; and went on to spend my allowance on rods, reels, and store-bought lures. During my high school years, living in cutover, regrown land in northern New Hampshire, I hunted deer, heard ruffed grouse drumming, and every morning looked out my bedroom window on the ever-changing spectacle of the north side of the Presidential Range. I learned about regenerating forests by finding stone walls and cellar holes in the woods and seeing the marks of old logging operations blazoned on the mountainside in a line-sharp difference of tree species.

Catholicism, science, and nature all offered ways of accepting the universe, complete with confessions of faith and inspirational literature. Catholicism's lessons started with a question-and-answer catechism and went on to comic books with moral dilemmas, tales of redemption, and stories of the martyrs. Science had books that told about great men (and Madame Curie), who were as dedicated as the saints and, like them, held to the path of truth—except here the path was science, not the commandments, and dying of disease or radiation lacked the drama of being thrown to the lions or roasted alive. Other books told about the wonders of the universe ("If the sun were the size of a grapefruit, then the earth would be . . . "). I read them all, but mainly the ones describing the world around me, at the intersection of science and nature. By the second grade I was deep into Thornton W. Burgess's tales of life in the Old Lily Pond and the Green Meadow, where Jolly, Round, Red Mr. Sun smiled down on Peter Cottontail, Reddy Fox, and Grandfather Frog. Ransacking the shelves as I grew older, I found Ernest Thompson Seton, Charles Roberts, and other nature writers who had inspired people like Aldo Leopold and Loren Eiseley a generation or two before (small town public libraries ran more to attic sweepings than new books). Only Catholicism had a formal theology, but science and nature had their own looser versions. Popular science described a material universe where science dispelled the darkness of ignorance with the light of reason. Nature had its own theology, though I had little contact with it in those years. In it, common sense and observation brought us into a realm of love in which we were truly at home. In works like John Burroughs's *Gospel of Nature* those sentiments attained religious force. One of my colleagues told me that an older relative pressed a copy of Burroughs's work into her hands with the explanation that this was as close to religion as anything she had ever found.

As a child, I did not care about how beliefs and experience fit together, or if they did, and no one pushed me on the subject. Popular science I took as a description of the world with no relation to religious belief, and the closest I came to the theology of nature was reading *Walden* my senior year in high school, a book E. B. White called the "best youth's companion yet written by an American."[3] I prefer the term "life raft." My catechisms described nature as God's creation, but did not oppose science, said nothing about Darwin, and failed to coach me in pious thinking in the woods. Nature was immediate, overwhelming, mysterious, and equally distant from the God of the catechism and the science of the textbooks.

Out in the woods, names and creeds vanished, leaving me with sensations and experiences that did not easily map back into formal knowledge. What had *Rhus toxicodendron* or the English words "poison ivy" to do with those shiny leaves? I understood the names, knew the classic warning ("Leaves three, let it be"), knew I was sensitive to poison ivy, but went through childhood summers covered with tannic acid, then calamine lotion, and then patent poison ivy remedies Mom found at the drugstore. In college I encountered people who believed that science explained everything and proved conventional religions were superstitions, and poking around on my own I discovered more approaches to the questions intellect raised about faith (and vice versa) than catechism classes had covered. Living, for the first time since my preschool years, with sidewalks and amid blocks of houses, I reflected on what nature meant to me.

Making meaning from information and experience, using and criticizing the systems I grew up with, was a messy process, for systems never quite fit life. The culture divided "science" from "religion," "knowledge" from "faith," and "reason" from "emotion." Science held that humans could understand the universe, while religions believed some things were beyond humans—"mysteries." Science saw "wonder," some mixture of astonishment and admiration, as our proper reaction to the beauties of the world, while creeds asked for "awe," reverence and a touch of dread, in the face of what was beyond human beings. But the neat divisions broke down. Science, as a matter of policy, denied mystery. Officially, it held, as definitely as a fundamentalist preacher cleaving to the Word of God, that human reason revealed all and that everything was only matter. In practice, scientists, particularly field biologists and physicists, smuggled mystery in the back door, for they had a sense of wonder about their subjects that shaded into awe. Science officially excluded talk of ultimates, but scientists used science to that end. Some followed Steven Weinberg, who declared that the "more the universe seems comprehensible, the more it also seems pointless"; others, like Ursula Goodenough, took the position that science showed we had a home in the world.[4]

Popular science magazines, presenting and interpreting the advance of science and technology, supported a vigorous theology that celebrated the wonders of a fully material universe and the human mind that could appreciate them.[5] Religious denominations—except those that denied human reason a role in understanding the world—used science, at least as a limit on theology (good theology could not deny what science had really found), and had, until Darwin, used science's wonders as evidence of the attributes of God. Darwin made that a harder but not impossible job. Michael Ruse, from science, gave a cautious "yes" to the question "Can a Darwinian be a Christian?", and John Haught, from theology, an enthusiastic one in *God After Darwin*.[6] Knowledge seemed, ultimately, to rest on faith in something, if only faith that we could know something, and faith, in turn, used knowledge to interpret revelation. Biblical fundamentalists and scientific materialists took absolute stands and defended clear and simple truths, but in the middle many rejoiced in a complex world and incomplete answers as the invitation to a lifelong search and spoke, often respectfully, to their opponents.

Environmentalism made the private business of reconciling systems an urgent and public matter, especially in the late 1960s and early 1970s, when it seemed that either the world's ecosystems or industrial society would collapse in the near future. Even when it became apparent that our situation was more a slow-motion disaster than a crisis, human numbers and technology still, in the long run, threatened the processes of life on earth, and environmentalists saw preserving them as a sacred trust. Dave

Foreman declared the preservation of wilderness was "an ethical and moral matter. A religious mandate."[7] Others made the same claim in more temperate language, and millions recycled their cans, bicycled to work, contributed to environmental causes, and put their retirement money into "green" funds. Environmentalism developed and persisted because it addressed great questions in the culture's authoritative terms, and because people found in its description of a threatened world and its call for us to take responsibility for the land a way to make a "proper connection with the higher powers."[8]

Since it comes down to the present, this book necessarily ends in speculation, but it begins in history, for it grew out of my convictions that environmentalism has roots in our culture and that only if we understand where it comes from can we understand what it is and think effectively about where it might go. The movement's concern with ultimate questions, it seems to me, did not come from the counterculture, Rachel Carson, or even Aldo Leopold, but from Western thought as it developed from the seventeenth century. For that reason, the first two chapters deal with what I see as environmentalism's dual legacy: the Enlightenment's rational approach to nature and Romanticism's concern with our emotional ties to it. Modern science has provided powerful and effective explanations that people can use to shape the world, but it has also given us a new way of understanding the world and a new description of what it is. While it is too much to say that science destroyed a world of mystery, wonder, and life and put in its place a dead world of particles in motion, it did reshape people's conceptions of the world.

By the early twentieth century, science and reason, not revelation and Scripture, formed the underpinning of public understanding and debate, and belief in the supernatural, which Western societies had regarded as the basis of civilization, became a matter of private opinion. Environmentalism relied on science, but it also embraced the contradictory legacy of Romanticism, which looked to intuition and emotion as guides to a "deeper" truth, and insisted on searching in the world for ultimate meanings. Emerson and his disciples established "nature" (which his disciples came to see as wild country) as the source of wisdom, refuge from society, and opening to reality. Environmentalism looked to science for facts and backing for its understanding of the environmental crisis, and to Romantic nature to guide a search for ultimate meaning.

Wilderness, the subject of chapter three, bridged the gap between Romanticism's individual appreciation of nature as part of life and environmentalism's wider concern with the relationship between our species and nature's systems, out of which we evolved and on which we still depend. It was the last pre-environmental nature campaign and one of environmentalism's first causes. The new movement helped make the wilderness journey a spiritual search, a pilgrimage combining outdoor recreation and the legacy of the pioneers with a quest for insight, and it made the protection of wilderness a sacred cause.

Chapters four and five take up environmentalism's development of religious values from the time it became a conscious and self-conscious movement in the late 1960s to the present. With *Silent Spring*, Rachel Carson crystallized Americans' vague fears about our world by tracing the spread of dangerous and possibly dangerous chemicals and proposing a radical and fundamental solution: that we view such apparently disconnected things as dirty air and water, pesticide residues in food, wildlife extinctions, and dwindling wilderness as symptoms of our mistaken belief that humans are the lords of creation. Carson called on us to abandon dreams of conquering nature and learn to live as plain citizens of the biotic community.

The uproar over *Silent Spring* helped environmentalism became a public cause, which, though it generated books and theories, developed more through actions than ideas. People campaigned for new laws, but they also put the brick in the toilet tank and looked for personal ties to the land around them. The more dedicated struck out to make environmentally responsible lives on the land, working out what it meant to live with and nourish nature in a particular place. The early enthusiasm faded by the late 1970s, but the movement continued. Environmentalists settled in for the long haul, looking beyond policies and programs for a marriage of nature and culture in a new society that would bring nature's graces into a new kind of everyday life.

The conclusion looks at the future, beyond predicting but of considerable interest. Analysis will not tell us how or if environmentalism is going to change society, but it can suggest what challenges the movement faces and how it might benefit from more consciously seeing its commitment as a way of accepting the universe. Environmentalists do not, generally, believe the movement constitutes a religion (and in conventional terms it does not), and they are uncomfortable with religious terms, but they ask religious questions: what purpose do humans have in the universe, and what must they do to fulfill it? Environmentalism's future seems to lie in the possibilities offered by its situation as a secular faith, searching for a transcendent within this world, building a morality on our scientific understanding of our place in it, and using science, which refuses to ask about purpose, to search for ultimate meanings. Facing that paradox, solving or embracing it, remains environmentalism's central task.

1 / Newton's Disciples

The environmental movement began with a concern for what was happening at the time—with DDT in our body fat and organochlorines in our drinking water and the awful environmental news of the late 1960s, the discovery that economic development and population growth threatened wilderness and even the world's ecosystems.[1] In the spring of 1970, when a quarter-million people rallied in Washington, D.C., for the first Earth Day and many more at teach-ins and protests across the country, they wanted to change the laws, not discuss humanity's deep ties to the world.

Politicians also focused on the present. They had passed laws to save parks and manage resources; now they passed laws mandating preservation of "the environment." The National Environmental Policy Act spoke in general terms, but most legislation had specific targets. Laws set targets for the reduction of pollution in the air and the water, banned some chemicals and regulated others, established a program to save endangered species, and imposed specific duties on industries and people. Like the environmentalists, the politicians spoke of the moment. President Richard Nixon declared that the 1970s were the decade in which we had to act: "It is literally now or never."[2] Many agreed.

A SENSE OF WONDER

Environmentalism, though, does not make sense without the deeper current of concern about the way humans are related to the world. Rachel

Carson attracted public attention and gained headlines with warnings of the dangers of pesticide residues, but achieved lasting popularity not from muckraking, but morality. She preached—that is not too strong a word—that we had a duty to life on earth, the "obligation to endure." We had to understand our place in the world, change our values and our hearts.[3]

Even at the time, people realized that Carson's case involved more than human health. Her admirers made her a modern nature saint, and her opponents looked more at her values than her science. Speakers for the National Agricultural Chemicals Association questioned her scientific credentials and evidence, but they charged with much more fervor that she opposed science, progress, and Western civilization, that if we took her ideas seriously we would soon be living in caves and eating nuts and berries. To be human, they said, we had to conquer nature, and modern society had written a glorious page in the annals of civilization by developing the wonderful chemicals that had saved so many people from disease and so much food from the ravenous hordes of our insect enemies. There was a lot of injured professional pride and economic interest in these accusations, but also outrage at what her opponents saw as Carson's denial of the power of human reason and her rejection of the triumph of the human spirit that produced modern civilization.

Carson's opponents were wrong in thinking the public would rally to the old cause—most people saw at least some merit in Carson's case—but right in seeing the contest as fundamental.[4] The "conquest of nature" and living as "a plain citizen of the biotic community" involved different ways of accepting the universe. One view denied that humans were part of nature or at home in it; the world was only raw material to be used for what we wanted. The other declared that the world was our home, that we were creatures of nature as well as culture who were charged with the duties of good citizens not just toward human society but the larger community of nature. We needed nature to live a full and fully human life.

Before she wrote *Silent Spring,* Carson's reputation rested on her writings about wild nature, and her own love of the earth and its life lay at the foundation of her activism. Like many other nature writers and environmentalists, her fascination with nature began early, aided by her

mother and life on sixty-five acres in Springdale, a small town east of Pitts-
burgh. Her academic interests, though, seemed entirely literary until her
junior year in college, when a required biology course led her into science. She received an M.S. in biology in 1932 and spent the next twenty
years as a government biologist, doing research and editing publications,
while pursuing a second career as a nature writer. Her first book, *Under
the Sea Wind* (1941), attracted little attention (probably because it appeared
about the time the Japanese attacked Pearl Harbor), but a second, *The Sea
Around Us* (1952), made the best-seller lists. Carson became and remained
a full-time writer until her death in 1965.

She argued for nature by implication in *Under the Sea Wind* and *The
Sea Around Us* and directly in an article for *Women's Home Companion*
in 1956, "Help Your Child to Wonder," which was later printed as a book.[5]
"A child's world," she said, "is fresh and new and beautiful, full of wonder and excitement. It is our misfortune that for most of us that clear-
eyed vision, that true instinct for what is beautiful and awe-inspiring, is
dimmed and even lost before we reach adulthood." She appealed to parents to give their children "a sense of wonder so indestructible that it would
last throughout life, as an unfailing antidote against the boredom and dis-
enchantments of later years, the sterile preoccupation with things that are
artificial, the alienation from the sources of our strength." "The lasting
pleasures of contact with the natural world are not reserved for scientists
but are available to anyone who will place himself under the influence of
earth, sea and sky and their amazing life." "Those who dwell, as scientists
or laymen, among the beauties and mysteries of the earth are never alone
or weary of life. Whatever the vexations or concerns of their personal lives,
their thoughts can find paths that lead to inner contentment and to
renewed excitement in living. Those who contemplate the beauty of the
earth find reserves of strength that will endure as long as life lasts. There
is symbolic as well as actual beauty in the migration of the birds, the ebb
and flow of the tides. . . . something infinitely healing in the repeated
refrains of nature—the assurance that dawn comes after night, and spring
after the winter."[6]

Here Carson touched on the heart of the environmental movement:
the passion for nature that fills many people's lives and touches many more.

What but an eagerness for this other world of nature would keep you rambling through wet woods on a cool spring day until not only your clothes but every piece of paper in your wallet was soaked? Why else take a child—past her bedtime—out under the eaves in a rainstorm to watch the lightning strike, or walk in the woods on a moonless midnight with no flashlight? What but the belief in a reality beyond human society but still part of this world sustains delight at the daily swoop of a barn swallow under the porch eaves or the quick glimpse of a mockingbird dropping two stories from house gutter to lawn?[7] Behind the public face of environmentalism—its scientific studies, calls for protection, programs, and regulations—lay personal connections and beliefs. Fears about the impacts human numbers and technology had on wild nature fueled the movement, but so did the belief that we were losing more than resources or beauty: we were losing something essential to the human spirit. A sense of physical crisis drove the movement's reform program—people worried that things were going to collapse, and soon—but a concern with humans' relation to the universe, which reached the level of religion, lay beneath the campaign for new laws.

Neither environmentalists nor their opponents saw environmentalism in religious terms because they did not think of religion in terms of ultimate questions but denominations and creeds. Religions were institutions that told about a world beyond the senses. Americans viewed their own beliefs in human reason, the autonomous individual, human independence from nature, and the supremacy of material values not as beliefs about our place in the world, but as undeniable facts. In the usual scheme the world was one thing and humans another, and we should shape the world to our needs. If everyone worked for their own interests, the Earthly Paradise of consumer goods would follow, and eventually, in the more enthusiastic versions, humans would spread to the stars. These beliefs were not, perhaps, deeply spiritual, and did not address some fundamental questions, but the idea of humans controlling the universe and the flood of goods and gadgets made for an attractive faith. Environmentalism denied that humans were all, that reason enabled us to build an Earthly Paradise of gadgets, and that dollars measured all values. For these heresies the right-minded condemned it.

BELIEF IN REASON

The triumphs of science and technology had allowed the conqueror's view by making it unnecessary to invoke Divine Intervention. Science's infrastructure of societies, museums, and publications seemed to provide for the irresistible and inevitable advance of human knowledge. James Watt's steam engine and later triumphs of technology provided visible demonstrations of human power over nature. Firearms defeated wild beasts and non-European peoples; mechanical power drained mines and took us across the land at unheard-of speeds. By the late nineteenth century the cultural center of gravity had shifted from revelation to reason.

In the twentieth century educated people commonly dismissed "religion" as folk belief, superstition, or a stage of cultural progress society had, happily, passed beyond—at any rate, something no one with any pretensions to knowledge or sophistication took seriously. Science became the only acceptable source of understanding; objective, reproducible data, the only true knowledge; and supernatural religions, which people had seen as the support of civilization and the foundation of society, private opinion. Ultimate questions and ultimate meanings ceased to have any public value. Science stopped its explanations at the level of instrumental values and material causes, and people followed its lead, abandoning— at least openly—the search for ultimates. The "infinitely healing in the repeated refrains of nature" and people's joy in the diversity of its life became simply phenomena, perhaps a source of private satisfaction and pleasure, but without significance for society.

Just because the culture made organized religion an optional part of life did not mean people ceased to believe; they just ceased to call their beliefs religion. Secular beliefs lacked the characteristics we associate with religious organization, but they functioned as religious creeds, describing the world and humanity's place in it. Though some strenuously denied it, these were systems of belief, for it could no more be demonstrated that humans could understand the universe than that a white-bearded God in a robe created it in six twenty-four days, no more be proved that our senses showed us everything than that after death the righteous would sit around on clouds and strum harps. Allied with nationalism and patriotism, the American version of modern secular belief even acquired many of religion's trappings. "The American way of life" included a daily ritual for schoolchildren (the pledge of allegiance to the flag), yearly remembrance of those fallen in battle, and, until recently, something that seemed a form of Absolute Evil: Godless Communism.

During the Cold War guardians of orthodoxy hauled heretics before the House Un-American Activities Committee to be shamed by congressmen. Confession and repentance before these high priests, of course, restored them to grace.[8] These public expressions of virtue did not address questions of ultimate purpose or such things as individual survival after death, but conventional faiths filled the gap by blessing the nation. Cardinal Spellman of New York, for example, assured American troops fighting in Vietnam that they were Christ's soldiers, and preachers emerged from the Oval Office to assure the public that the president was right with God—or, later, in the case of President Clinton, had at least wrestled with his demons. This was a faith that dared not speak its name, but it did not need to. As the common public point of view, it escaped scrutiny.

Environmentalism developed within the modern world, seeing secular beliefs as all there was, the material world as the only one, and human reason as our only tool for understanding, but it rejected other common views. Environmentalism believed nature had an intrinsic value and rights that humans should respect and that some values could not be reduced to dollars. It looked beyond knowledge, seeking meaning, and believed each of us needed to form conscious ties to the world. These concerns led some to see the movement as irrational or "mystical," a weak-minded search for emotional satisfaction, or the refuge of those who had lost touch with reality. Those views had little foundation. Some environmentalists were mystical or weak-minded or out of touch with reality, as were some stockbrokers and free-enterprise Republican congressmen, but in its conscious attention to humans' place in the world, its insistence on nature and non-monetary values, environmentalism addressed questions fundamental to human life and looked steadily at real problems.

Its focus on choices and action was more "realistic" than belief in an inevitable progress that would solve our environmental problems and economic growth as a complete measure of public welfare. Besides, the anti-

environmental opposition, despite its appeals to statistics and human reason, had as much ideological and philosophical baggage as any starry-eyed
tree hugger and resorted to emotional and "irrational" arguments as commonly as New Age gurus. Anti-environmental arguments appeared hardheaded only because everyone accepted paeans to Progress and sermons
on the Questing Spirit of Mankind as self-evident truths.

Looking at environmentalism as an approach to ultimate questions
takes us back to questions of what we believe and why we believe it, for
society's generally accepted view forms the ground on which environmentalism was built and the beliefs it holds to and departs from. This chapter touches upon the main points of modern secular beliefs. It begins with
the rise of the Enlightenment view within seventeenth-century Europe
and its triumph throughout Western society during the next two hundred years, then takes up Romanticism's role in making the conquest of
nature a heroic venture and America's destiny. Americanism, the complex of commerce, patriotism, and piety that saw Reason, the Hand of
God, and the Market leading humanity into a future of endless Progress,
provided the context for environmentalism. Environmentalism rejected
parts of the conventional wisdom but accepted much of its foundations.
It believed, for instance, in science and had as much faith in reason and
laws as any conventional businessman. Environmentalism, like environmentalists, faced the challenge of fusing these ingredients—science's
objective stance, a Romantic attachment to experience in nature, and the
conditions of modern industrial life—into a satisfying vision of human
life and purpose that took account of environmental realities.

FAITH IN SCIENCE

To describe the Enlightenment as a "recovery of nerve," as one historian
did, understates the case.[9] The seventeenth-century intellectuals (usually
known by the French term, *philosophes*) who saw reason as humans' only—
and sufficient—tool to understand the universe, established ideas about
humans and their place in nature that still guide our hopes. The obvious triumphs of the mind in seventeenth-century Europe fueled their
dreams. Copernicus's astronomy showed the sun, not the earth, at the cen-

ter of the solar system, and Galileo's telescope revealed new wonders in the universe: craters on the moon, rings around Saturn, and moons orbiting Jupiter. Newton's physics allowed humans to calculate the fall of an apple and the movement of the moon and to see that the same forces controlled both. The *philosophes* made Newton the great lawgiver: "Nature and Nature's laws lay hid in night/ God said, Let Newton be! and all was light."[10] Elevated to a philosophy, the methods and assumptions of physics destroyed a cosmology, reduced Creator to clockmaker, and changed Western views of the world. Instead of being God's mysterious handiwork, held in existence by his immediate and continuing intervention, the heavens and the earth appeared as a complex of bodies and forces governed by laws we could grasp.

Confidence in reason's power grew through the eighteenth century. In Europe Descartes's emphasis on rigor and reason and his view of geometry as the model of rational thought shaped the thought of a generation (and in France, the society) and in Britain Francis Bacon's program for a new science based on observation and experiment that would allow humans to understand and master nature formed the idea of science. The educated minority, at least, debated plans for a rational social order grounded in principles revealed by reason and believed people could codify and arrange all knowledge. Those dreams persist in the belief that science will solve our problems, in plans like E. O. Wilson's—developed in *Consilience*—to take up again the Enlightenment project of unifying all knowledge and even in our government, for it rests on Enlightenment principles. The Declaration of Independence, with its emphasis on "Nature and Nature's laws," its belief that each people could and should arrange their government as seems best to them, and its appeal to the opinion of mankind, rested on faith in reason. So did the idea of a Constitution. The *philosophes,* in turn, saw the American experiment as a test of their theories and evidence of human progress. The confidence in reason that produced the eighteenth-century *Encyclopédie* has its modern equivalents, and the Enlightenment program of a unified, universal knowledge remains a compelling dream.

Naturalists (the term "scientist" appeared only in the early nineteenth century) probed what had been nature's mysteries with increasing success. In the mid-eighteenth century a Swedish botanist, Carolus Linnaeus, developed a system that organized the flood of specimens returning voyagers brought to Europe, giving us not only modern taxonomy but the Latin binomial names people learn in high school biology. In the early nineteenth century Charles Lyell explained changes in the earth's surface as the result of purely physical forces, the same ones we see at work today, acting over long periods of time.

Despite its challenge to a literal interpretation of the earth's creation in Genesis, Lyell's theories hardly changed religious ideas at the time. They did impress the young Charles Darwin, who took Lyell's first volume along on the *Beagle* and tried to get the others as they came out and he voyaged around the world. Darwin went on to crown natural history's triumphs by showing how environmental pressures on the apparently trivial variations among individual plants and animals could, over the ages, produce life in all its dizzying variety.[11] With no need to call on divine or even conscious intervention in life and development, atheism became intellectually respectable.

Knowledge continued to grow, and with it power. "Knowest thou the time when the wild goats of the rock bring forth?" God mockingly asked Job. We can look that up in *Walker's Mammals of the World*. "Doth the hawk fly by thy wisdom?" (Job 39:1, 26). No, but we know how it turns the trick, and since we have shotguns, it flies at our sufferance.[12] Technology gave visible evidence of human power as well as human understanding. Almost as soon as tracks were laid, the railroad became the icon of Progress, particularly in the United States, where the railroad allowed nineteenth-century Americans to bring nature to market as buffalo hides, grain, lumber, and ore. Lithographs commonly showed the locomotive, black smoke pouring from its stack, moving relentlessly into the wilderness, sweeping away forests, Indians, and gloomy skies, spreading in its wake fertile fields, solid farmhouses, and great cities under a shining sun. The telegraph, and then the submarine telegraph, sent news around the globe in an instant. Civil engineering—not spectacular, but necessary—allowed cities, for the first time in history, to grow by natural increase rather than draining the countryside to make good the toll from disease. Medicine, linked to science in the late nineteenth century, made many of the common diseases and early deaths uncommon or rare. People could believe that they were not at nature's mercy, even that nature was at ours, and Western societies—with the United States in the lead—came to see the conquest of nature as the destiny and great task of civilization and its crowning glory.

In the late nineteenth century, in a contest Andrew Dickson White called the "warfare of science with theology in Christendom," science replaced Scripture as authoritative knowledge about the world.[13] The new myth of reason and science triumphant began by telling about the persecution of Galileo for saying the earth went around the sun, which it pictured as a confrontation of the new knowledge with entrenched and corrupt religion (for in nineteenth-century Protestant America nothing evoked superstition and intellectual tyranny so well as the Catholic Church). The struggle of Lyellian geology against the literal reading of Genesis and the historical reality of Noah's Flood provided the second act. The climax came in the battle over Darwinian evolution, represented by the debate between Thomas Huxley, Darwin's great defender, and Bishop Wilberforce, holding on for a literal reading of Genesis. In the standard retelling, science, representing Truth and Reason, liberated humanity from the shackles of ignorance and the superstitions of theology.[14]

The myth formed and informed opinion and shaped public debate through the twentieth century. In 1925 the town fathers of Dayton, Tennessee, restaged the clash by prosecuting John Scopes for teaching evolution in the local high school. Clarence Darrow (famous lawyer and skeptic) stepped in to defend Scopes, and William Jennings Bryan (equally famous populist orator, fundamentalist Christian, and national politician) came to the prosecution's aid. The trial was a news sensation, then a book, a play, and a movie *(Inherit the Wind),* and a rhetorical reference point and touchstone for the battles over "creation science" in American public schools that still go on.

The standard account made science and theology inevitable opponents and science (with Truth on its side) the inevitable victor, but none of that was true. Theologians had always used authoritative secular knowledge, even if they had not completely agreed with it. Early Christians interpreted the Gospel in terms of classical civilization's categories and concepts, and

in the thirteenth century Thomas Aquinas became a saint for reconciling Christian doctrine and Aristotle. He was not alone; the great medieval
Jewish scholar Maimonides devoted considerable effort to establishing the
relation of secular philosophy to God's commandments. All their work
began with the belief that while only God could know everything about
the universe, humans could know something; the problem was not,
therefore, to choose either secular or sacred knowledge, but to draw the
boundary between them in the right place.

Even in the heat of the Darwinian debates, some scientists and theologians held to this middle position, that faith and reason were not necessarily opposed; today, even as Richard Dawkins, Daniel Dennett, E. O. Wilson,
and Stephen Jay Gould argue that science is all we need, others—
scientists, theologians, and citizens—use Darwinian ideas in support of
conventional religious stands.[15] The myth of opposition remained popular, though, for it had enough truth to make a good story, and its neat
division of right from wrong and saved from unsaved served fundamentalists and scientific materialists alike by giving an air of heroism to those
upholding the true faith against (choose one) the Godless secular human-
ists or the forces of ignorance and reaction.

Accepting science as authoritative knowledge about the world, though,
did more than substitute engineering for epistemology. Encouraging
people to see the world as only matter and energy, science looked toward
what nature writer Annie Dillard called "the unhinging of materials from
meaning."[16] Ending mystery, science abolished the sacred, a point of view
that Mirceau Eliade, a philosopher of religion, believed would ultimately
not satisfy human needs. "Experience of a radically desacralized nature
is a recent discovery; moreover, it is an experience accessible only to a
minority in modern societies, especially to scientists. For others, nature
still exhibits a charm, a mystery, a majesty in which it is possible to deci-
pher traces of ancient religious values."[17] Eliade's view that the Scientific
Revolution turned a living world into dead matter and separated us from
the sources of life is a common one. Environmentalists, for example, often
look to primary cultures for examples of respect for nature and ways to
incorporate nature into the culture At least one anthropologist believes
these groups are no more religious or respectful of nature than we are.

Whatever the case, many in the modern world refuse to see humans as detached consciousness in a universe of mechanical laws or as something evolution produced to appreciate its marvels. They look to nature for "a charm, a mystery, a majesty" they need in their lives.[18]

For many, though, belief in the supernatural came to seem unnecessary, then foolish. To them, what we saw did not conceal a greater reality— what we saw was all there was. We could not strike through the mask and get beyond appearances to a higher reality. The mask was all there was. In Western societies religion ceased to be the ground of reality on which society depended and civilizations were built. It became a private belief about something for which there was no evidence. Public figures paid lip service to the pieties of established faiths (as aids to social order) but justified action by material measures. Life found a new meaning as well. In a world of supernatural power and revelation, life had been a mystery to be lived, and what meaning we could find was found only by living. Life became a problem to be solved and its meaning a matter of gathering information. Supernatural views commonly gave each person (and in most formulations, the human species) purpose and destiny, a task to fulfill and a relation to the universe, while the new view seemed to show the universe as only matter, humans as only organisms thrown up by chance operating through mindless processes, and our only tasks and purposes ones we found ourselves.

Because people rejected the supernatural did not mean they gave up searching for order in the world and a purpose for our species. They only looked for it outside what they saw as "religion." E. O. Wilson's movement from Christianity to reason shows that shift. Raised in the family's Southern Baptist faith, he had, by his college years, shifted his devotion to natural history. "I was," he said, "enchanted with science as a means of explaining the physical world, which increasingly seemed to me to be the complete world. In essence, I still longed for grace, but rooted solidly on Earth." He rooted it by arguing that evolution gave us an attraction to nature and even to particular kinds of nature, an attraction to which he gave the name "biophilia" and discussed in a book of that name.[19] Humans prefer, for instance, open scenery like the African plains where the species evolved and are wary of snakes even without experience of them. Modern biology gave that fascination a new focus, for it "produced a genuinely new way of looking at the world that is incidentally congenial to the inner direction of biophilia." The journey of discovery is an endless quest that "will engage more of the things close to the human heart and spirit. It seemed possible that the naturalist's vision is only a specialized product of a biophilic instinct shared by all, that it can be elaborated to benefit more and more people. Humanity is exalted not because we are so far above other living creatures, but because knowing them well elevates the very concept of life." Understanding organisms leads to "placing a greater value on the world and ourselves." The great hope for the conservation movement lay in "the understanding of motivation, the ultimate reasons why people care about one thing and not another. . . . The goal is to join emotion with the rational analysis of emotion in order to create a deeper and more enduring conservation ethic."[20]

Much of Wilson's writing looks toward that goal, including his recent *The Future of Life* (2002), which begins with a letter to Thoreau. Here he follows a long line of naturalists arguing, as Rachel Carson had, that contact with nature and a "sense of wonder" enrich our lives, a line that included not just scientists but educators as well.[21] Nature study, developed as an elementary school subject in the first decade of the twentieth century, encouraged children's sense of wonder as a way to make rural life better and richer. The elementary school natural history pamphlets and readers built on that with gosh-wow statistics. Learning that the weight of insects in the world exceeds the weight of mammals startled children, but it also put the confusion of grasshoppers, crickets, ladybugs, praying mantises, and beetles out there in the grass in a new light. At a later age the knowledge that hummingbirds at the backyard feeder in spring and fall are part of an annual ebb and flow from the tropics to the Arctic, an avian cycle constant since the melting of the Pleistocene ice, added something to the simple pleasure of seeing them. Knowing that the atoms in our bodies come from dead stars and will nourish the world after we are gone links us to the cosmos and to the round of life.

Scientific materialism (the belief that human reason suffices to explain a world that is wholly material) relied on science but rested on faith. It was one thing to say humans were the result of evolution, another to say that was all they were. Darwin and the geologists made it impossible to hold to both a literal interpretation of the Bible and modern science, but rejecting a literal interpretation of the Bible did not—except in the fundamentalist view—mean rejecting the Bible. Even in the nineteenth century many saw evolution as a mechanism, not a final cause. Scientists, as scientists, usually rejected the whole idea of final causes (ultimate purposes)—as Steven Hawking put it, he did not deal with "God questions"—but that was itself a religious position, a stand on ultimate questions.[22] That humans could understand the world was also an article of faith, one even some scientists rejected. The British population geneticist J.B.S. Haldane famously remarked that he had come to suspect "that the universe is not only queerer than we suppose, but queerer than we *can* suppose."[23] Recent work on quarks and superstring suggests Haldane did not fully appreciate human abilities to suppose what was queer, but should we dismiss him? If we wished to, what grounds would we have? Even E. O. Wilson, who argued that the *philosophes* were right in pinning their hopes on human intellect and saw science as, ultimately, explaining everything, used many "ifs" and conditionals in his argument, and admitted he might be wrong.[24]

Scientists commonly denied science was a religion by claiming humility in the face of nature and pledging their allegiance to "facts." Carl Sagan declared in *The Demon-Haunted World: Science as a Candle in the Dark* that he did not worship "at the altar of science. . . . The directly observed success of science is the reason I advocate its use" (p. 30). Scientists and science popularizers, though, often show what Mark Twain called the calm confidence of a Christian holding four aces, and the idea of science as a "secular religion for a new age" ran through popular expositions of science from the nineteenth century.[25] Sagan himself celebrated science as our guide to ultimate understanding—our only guide.

In *Cosmos* (a television series as well as a book, its title echoing the magisterial synthesis of the great naturalist Alexander von Humboldt, *Cosmos: A Sketch of the Physical Description of the Universe*), Sagan declared his faith in the materialist conception—"The Cosmos is all that is or ever was or ever will be"—and he found no value or truth in beliefs about the transcendent. "For thousands of years humans were oppressed—as some of us still are—by the notion that the universe is a marionette whose strings are pulled by a god or gods, unseen and inscrutable. Then, some 2500 years ago, there was a glorious awakening in Ionia. . . . Suddenly there were people who believed that everything was made of atoms; that human beings and other animals had sprung from simpler forms; that diseases were not caused by demons or the gods; that the Earth was only a planet going around the Sun. And that the stars were very far away."[26]

Sagan measured conventional religions against science and found them wanting. That "so little of the findings of modern science is pre-figured in Scripture to my mind casts further doubt on its divine inspiration." We should trust the guidance of science, Sagan said, for it created wealth and would save us from the problems caused by technology. It gave us our "deepest" knowledge of the universe, and, if not the same thing as democracy, strongly supported it. Science was even spiritual. There is "no necessary implication in the word 'spiritual' that we are talking of anything other than matter. . . . Science is not only compatible with spirituality, it is a profound source of spirituality. Its description of "an immensity of light years and . . . the passage of ages . . . is surely spiritual."[27] This, as someone once remarked in another context, was the Truth that would make you free.

Sagan's hopes appear in less academic form in science fiction about immortal minds in robot bodies and in the business of freezing bodies for revival when we can cure all diseases, but there are sober and philosophically sophisticated versions as well. In *Consilience* E. O. Wilson calls for a revival of the Enlightenment program: "The assumptions [the Enlightenment philosophers] made of a lawful material world, the intrinsic unity of knowledge, and the potential of indefinite human progress are the ones we still take most readily into our hearts, suffer without, and find maximally rewarding through intellectual advance." The unification of all knowledge should stir our blood. Its "strongest appeal . . . is the prospect of intellectual adventure and, given even modest success, the value of understanding the human condition with a higher degree of certainty."[28]

According to Wilson, science rests on "the faith that if we dream, press to discover, explain, and dream again, thereby plunging repeatedly into new terrain, the world will somehow come clearer and we will grasp the true strangeness of the universe."[29] We are "obliged by the deepest drives of the human spirit to make ourselves more than animated dust, and we must have a story to tell about where we came from, and why we are here." The Bible may have been an early attempt to do that, and "perhaps science is a continuation of new and better-tested ground to attain the same end. If so, then in that sense science was religion liberated and writ large. Preferring a search for objective reality over revelation was another way of satisfying religious hunger," an old quest, "intertwined with traditional religion," but on a different course. It aimed "to save the spirit, not by surrender but by liberation of the human mind. . . . Its central tenet, as Einstein knew, is the unification of knowledge. When we have unified enough certain knowledge, we will understand who we are and why we are here."[30]

There are drawbacks to the program *Consilience* argues for. Most notably, it gives only the scientists something to do.[31] The rest of us must stand on the sidelines and watch while science moves into the realms of the genome and the cosmos, hardly taking note of the plants and animals we find nature's most fascinating things. Worse, modern science presents a bleak universe. As Steven Weinberg, a physicist, put it: the "more the universe seems comprehensible the more it also seems pointless." He endorsed Wilson's program, but there is little to hold us in his view that the "effort to understand the universe is one of the few things that lifts human life a little above the level of farce, and gives it some of the grace of tragedy."[32] Only slightly better was the argument that without God or gods humans had the responsibility to choose wisely and to make ourselves in freedom.[33] That had a brave ring to it, but seeing human purpose as making a purpose seemed a bit too self-involved.

Primary cultures rooted their people in the world with myth and ritual, and even made people responsible for renewing the processes of life and keeping the universe in balance. Making meaning in an impersonal world offered nothing so appealing. Biologist Ursula Goodenough met the feeling of helplessness that came with Weinberg's pronouncement with a classical remedy. She accepted the universe "as the locus of Mystery," asking why there was a universe, where the laws of physics came from, and why the universe seemed so strange. "Mystery. Inherently pointless, inherently shrouded in its own absence. . . . Mystery generates wonder, and wonder generates awe. The gasp can terrify or the gasp can emancipate. As I allow myself to experience cosmic and quantum Mystery, I join the saints and the visionaries in their experience of what is called the Divine."[34] Mysticism, though, had little appeal for most people. They wanted something to do.

ENDLESS PROGRESS

Most people held to the more comfortable faith that the conquest of nature would make life easier for all. Politicians promised that science (which they usually confused with technology) would bring prosperity, and the popular press, from *Scientific American* to *Popular Mechanics,* hymned science and technology as the path to heaven on earth. Science fiction, which always had less to say about the future than about present dreams and nightmare, insisted on Progress from its Victorian birth. The pulp magazines that dominated science fiction from the 1920s discarded even the undercurrent of pessimism that marked H. G. Wells's early work in favor of science triumphant across the universe. The conquest of the universe was man's destiny (feminists in science fiction had an uphill pull). That idea held to the present, as witness Kim Stanley Robinson's recent Mars trilogy that told of humans transforming that planet from a barren desert into another Earth. It ends as humans prepare to launch the first ship to explore the galaxy. Robinson knew about ecology and sympathetically depicted opponents of terra forming, but he made them only figures for dramatic tension in the story of human Progress, and Mars's only value was first as an arena for human struggle, then as an ideal (because shaped to our order) playground. The trilogy's theme was that we would become as gods.

That kind of confidence formed the first line of defense against environmentalists' arguments that nature had limits humans had to respect, and the argument even appealed to many environmentalists.[35] The *Whole Earth Catalog,* which peddled environmentalism in its philosophies-for-sale section, declared in its epigraph: "We are as gods and might as well get good at it."[36] Even now there is a trickle of environmentalists repenting and embracing again the old-time religion of Progress. In *No Turning Back* (1994), Wallace Kaufman tells of abandoning what he sees as a false philosophy and distorted view of humanity. Humans have done well, he said, despite mistakes. We have "made the world an ever more livable place for ourselves by using science and technology and exercising our powers of dominion." Far from producing disaster, these have led to a better life for all. The future is not dark, Kaufman insists, "a new Renaissance in human understanding has already begun."[37]

Gregg Easterbrook, a journalist, made much the same case in *A Moment on the Earth* (1995), and the *Economist* and the *Wall Street Journal* joyfully hailed Bjorn Lomborg, former Greenpeace member, who returned to the fold with *The Skeptical Environmentalist* (2001).[38] Dedicated to Julian Simon, it argues (with statistics and 2,930 footnotes) that all indicators of human welfare point to a long-term, continuing rise in life expectancy, food supply, education, and health. We should, Lomborg concludes, be able to continue this progress indefinitely without destroying the global environment.

These arguments, Lomborg's in particular, ratified E. F. Schumacher's view that economics so dominated public discourse in modern society that to call something "uneconomic" was to damn it utterly. That was especially true in the United States,[39] where people commonly saw society as a collection of individuals, calculating and accumulating, their individual drives making for progress, and the community as the result of these individuals choosing to sacrifice part of their autonomy for the benefits of association. The market seemed from this point of view a natural force that governed all social interactions outside the family, rewarding hard work (the central virtue) with wealth while punishing lack of initiative (the great vice) with poverty.

This worldview shaped policy in areas far removed from the market, encouraging the view that the best (or the only) way to reduce crime was to increase prison sentences so much that criminals would decide the costs of crime were higher than its benefits and the idea that almost any social institution, including public schools, could be improved by the imposition of "market discipline." Politicians told of the genius of capitalism freeing human ingenuity, producing endless economic growth, which in turn freed us, as individuals and a society, from all limits. There was no need to restrict consumption or to redistribute income. Indeed, these were antisocial acts, refusals to join in enriching everyone. The Earthly Paradise of consumer goods politicians saw beyond the horizon was distinguished from the supernatural paradise only by being material and requiring neither prayer nor penance (and Republican versions did not even require taxes).

A romantic view of economic activity infused color into this gray view of isolated consciousnesses calculating their rational self-interest by associating business with risk-taking and freedom. We usually think of Romanticism as a matter of beauty rather than money and of nature rather than society, but as Isaiah Berlin pointed out in "The Romantic Revolution," that movement looked to nature, but it also made "the essence of man self-mastery—the conscious choice of his own ends and form of life."[40] Seeing the divine spark in humans not in reason but in creation, Romanticism made freedom the absence of restraint—liberation even from society's moral laws, for these inevitably clashed with the individual's realization of self.[41]

AMERICAN BELIEF

Americans agreed, taking (or mistaking) their cue from Ralph Waldo Emerson, who preached in "Self-Reliance" that "whoso would be a man must be a non-conformist" and declared that "society everywhere is in conspiracy against the manhood of every one of its members."[42] In Europe the key Romantic figure was the artist, true to himself and scornful of a society unable to understand his (it was always *his*) genius, but in the United States it was everyman, and in particular the businessman, making an enterprise stamped with his name and built by his decisions.[43] From this perspective, which was as Romantic as the more usual one, Nature appeared not as dame or mistress, guide, despot, or housekeeper, but as inert matter on which we worked our will and so showed our individuality.[44] It was only matter and had, so to speak, no rights that a white man was bound to respect.

Nineteenth-century Americans made the romance of enterprise a national dream, and the conquest of the West, where the dream was to be lived, the center of a national myth. "Westward the course of empire makes its way," led by the buckskin-clad pioneer, then the railroad locomotive. Government freed individuals from economic dependence by throwing open the gates of the endless storehouse of nature's riches, the public domain, and when the land proved not to be endless, Americans found new frontiers and new ways to work the old ones. Gifford Pinchot, the great apostle of conservation, strenuously denied that it meant "locking up" resources. Conservation, by which Pinchot meant national resource policies (saving scenery in national parks lay outside his mandate), stood for development, albeit more measured and controlled than the great land rush of the last two generations—good because it guaranteed opportunity for the individual rather than the great corporation and because, wielding the tools of science and rational organization, it would yield more prosperity for more people over more years than the scramble for riches that had already depleted our forests. Even efforts to save nature for its beauty made their peace with Progress. John Muir, the great apostle of national parks and wild nature, urged that we set aside beautiful areas as temples of Nature. Development would take place around them, and people could worship at Nature's shrines and refresh themselves at its fountains—to return with more energy to the economic fray.

Americans looked abroad for new commercial frontiers and to science for new frontiers within. The report recommending the establishment of the National Science Foundation after World War II described science as "the endless frontier." Space become "the final frontier" for both the American space program and the television series *Star Trek*, and today organizations like the L 5 Society and the Artemis Project urge us to embark on the great adventure of exploring the entire universe, a voyage they believe can begin in our lifetimes if only we trust the magic of the marketplace and the power of human reason.[45]

Independent individuals wresting wealth from nature and in the process enriching society formed the basis of the country's creed and civic religion, the "American way of life," which politicians preached and schools taught. This cult had its saints—businessmen, inventors, pioneers, and the heroes who defended it in battle—and an inspirational literature running from Benjamin Franklin's autobiography through the Horatio Alger books to the business and inspirational books on the shelves of the modern chain bookstore. These secular equivalents of Victorian Sunday school tracts presented models of striving entrepreneurs raising themselves by hard work from poverty or middle management (which recent literature viewed as the same thing) to success. With the fervor of a Methodist choir hymning the joys of heaven, everything from learned economic treatises to advertisements praised the wonders of the material goods that rewarded faithful attention to business. If Americans were not so determined to see their national values as secular, the catalogs that pour into modern middle-class households would peddle small golden calves suitable for home altars.

Americanism did not appear to be a religion because it denied it was one and because it called in conventional faiths as supplements to its beliefs. Presidents encouraged people to worship at the church or synagogue of their choice and on official occasions called on the clergy—until recently always Christian and almost always Protestant—to invoke God's blessing on their actions. Americanism, though, is a faith. It describes the world, humans, and the path of virtue. It sees Americans as a chosen people and gives them the mission of spreading democracy and freedom throughout the earth. It identifies the forces of evil (Godless Communism chief among them), presents the sainthood of money and power, and describes the ordinary virtues of those figures of ordinary success: the careful consumer, the friendly neighbor, and the good parent.

Americanism constitutes a religious view of life, but an incomplete one. All our philosophical or religious systems of belief necessarily fail to describe life, but some have more holes than others. Americanism promises some things, such as autonomy and power, but leaves to conventionally defined faiths other, equally important, questions. In exulting autonomy it neglects community, denying that people have vital connections to each other or a part in the workings of the universe. It does not address the issue of survival after death, something even so militant an atheist as George Orwell found a problem. The "decay of the belief in personal immortality," he said, "has been as important as the rise of machine civilisation." Its loss "has left a big hole, and . . . we ought to take notice of the fact." Elsewhere he called the decline in a belief in immortality the "major problem of our time."[46]

Americanism's emphasis on progress and its relentless optimism offered little protection against the tragic dimensions of human life. Rejecting even the possibility of failure, it saw setbacks, even defeats, as temporary and victory as inevitable, saw no limits, and believed every problem had a solution. Its stand inspired great dreams and encouraged dedication, but gave the culture and individuals no resources to grapple with intractable or even protracted problems. Crises it could deal with; chronic problems it could only ignore. Worse, its view made it difficult for people even to conceive of, much less address, a clash between two things each good in itself but incompatible with the other, deal with the shadows of defeat that are part of our lives, or face the reality of our deaths.

ENVIRONMENTALISM AND MODERN BELIEF

Environmentalism became a movement in the late 1960s from within the modern system of belief, but also against it. It accepted science as the path to knowledge, but rejected the ethos of "objectivity" and detachment and the picture of a purposeless universe often associated with it. It saw modern technology as necessary for our civilization but the "conquest of nature" that often accompanied it as a destructive mistake. It accepted America as a nation formed in nature, but held—possibly its most radical departure from orthodoxy—that nature had value apart from its use to humans and had the right to exist.

What made disagreements over values more than philosophical differences was the evidence that human numbers and technology now threatened to destroy many of the biological systems of the earth, systems on which humans, as well as wild nature, depended. The environmental movement developed in the atmosphere of crisis that accompanied growing consciousness of these problems, and it called for drastic changes not just in the way we lived but the way we saw ourselves and the world. It called on humans to limit their use of nature, in order to preserve other things, necessary in the long run for everyone. Despite the movement's radical foundations, though, environmentalists, like their opponents, looked for easy answers and rapid change and hoped to find a path that allowed us to have it all—consumer goods, technology, modern communications, wilderness, and a life in contact with nature—and they shrank from looking closely at the sacrifices an environmentally responsible society might require.

When environmentalism became popular, however, deep questions seemed beside the point. Environmentalism rested on people's revulsion at the industrial destruction of nature, but it was immediate concerns that turned vague dissatisfaction into a political movement and social program. Like the prospect of being hanged in the morning, radioactive fallout in milk and pesticide residues in body fat focused people's attention, and they demanded action that would help nature now. Endangered species legislation, begun to save a few visible emblems of wild America—the whooping crane, wolf, and bald eagle—by forbidding people to kill them, quickly evolved to cover all forms of life and activities that disturbed as well as killed species in peril. New legislation cut air and water pollution, regulated production, use, and disposal of a host of chemicals, and reduced the damage mining, lumbering, ranching, and farming did to the soil, plants, and animals. The social climate also encouraged a focus on the short term. The 1960s was a time of change and upheaval, with prophets on half the street corners and a widespread belief that "the revolution" was just around the corner. Every week the news had another cause and there were more new spiritual quests and gurus than new models of cars; many of them used ecology's vocabulary or saw it as part of their cause.

But while events drove environmental action, a deeper concern shaped its ends. Rachel Carson condemned careless use of pesticides but spoke with greater force against the attitude of mind that encouraged us to rely on them.[47] People read Aldo Leopold's work in part for its descriptions of nature and the delights and dilemmas of one who could not live without wild things, but they made him a nature saint for his argument about humans' moral duties toward the world. Environmentalism campaigned for human health, natural beauty, and a sustainable economy, but also insisted that nature was more than an amenity or even the spiritual necessity of a few who could not live without wild things—it was the ground of our lives. Rather than conquerors, we were plain citizens of the biological community and should behave that way. Environmentalism took its stand in ecology but also in the established, if minority, part of Western thought that examined humans' moral duties to the world that went back to classical civilization but had its modern roots in conservation.

In *Man and Nature* (1864), the first lengthy discussion of humans as a geological and biological force, George Perkins Marsh argued that man might use the world but should not waste its resources.[48] In *The Holy Earth* (1915) Liberty Hyde Bailey spoke of our duty to take "care that we do not despoil [the earth], and [warned we should be] mindful of our relations to all beings that live on it." We had, he warned, seen our relations to the land in terms drawn from "the realm of trade." We had to put them into "the realm of moral. . . . The morals of land management are more important than the economics of land management . . . [and] any line of development founded on accountant economics alone will fail."[49] Aldo Leopold made the same argument in *A Sand County Almanac,* though he relied on ecology rather than the Bible.

Environmentalists found they had to adapt these arguments, for in the late 1960s many people saw both science and religion as agents that had changed Western civilization's view of nature from a living world in which humans had a part to one that was only dead matter, a shift that paved the way for what they saw as the uniquely ruthless Western exploitation of the world.[50] The argument really began with Lynn White's essay, "The Historic Roots of our Ecologic Crisis," which shaped environmental debate for at least a decade and affects it to this day.[51] White saw Christianity, in particular Western Christianity, as separating matter and spirit and so creating the conditions that allowed the Western destruction of the natural world. Against the West he placed the example of Eastern religions, which encouraged humans to respect and live with nature.[52] Others looked to primary cultures and still others saw science as the decisive influence on Western attitudes.

Though popular, these ideas had their critics. Mary Douglas, an anthropologist, denied the neat division between modern culture and earlier ones and asked how long her colleagues were going to prop up "the popular piety" that modern peoples were "secular, skeptical and frankly tending more and more away from religious belief," while primitive peoples were "and always have been religious." Anthropological studies showed "plenty of secular savages. Indeed, in certain tribal places there is a notable lack of interest in the supernatural. God is not dead with western civilization. Science has not delivered the *coup de grace.*"[53] Others pointed out that White's dichotomy between Eastern religions that respected nature and a Christianity detached from and contemptuous of it was too simple.[54] Eastern traditions had not stopped Asian countries from using modern technology to pollute and destroy and Christian tradition had a strong strain of respect for Creation.[55] A more general criticism was that White's argument assumed that ideas were the mainspring of action, ignoring the possibility that people might do things first and then find ideas to justify them later.

The same points applied, with even greater force, to the argument that science had somehow destroyed spirit in nature. Into the nineteenth century, natural history had close ties to natural theology, and while Darwin made it difficult to find God's goodness in the smiling meadow, other things, from the daily wonders of spider webs to the longer ones of galaxies, remained a source of awe and wonder.[56] Scientists, in fact, particularly field biologists, often seemed fascinated with their subjects. As Edgar Anderson, a prominent plant taxonomist, pointed out, his colleagues, though they seemed exemplars of "objective" and imperial science as they collected plants for museums, were really "incurable romantics," deeply in love with the wild.[57] Stephen Jay Gould, Carl Sagan, and E. O. Wilson showed the wonders of science as the emotional foundations of a commitment to nature.[58] Whatever the intellectual implications of Cartesian geometry or Francis Bacon's method of questioning nature, scientific programs and popular science included as much love of nature and identification with it as they did conquest and reshaping of brute matter.

By the 1890s nature literature made the Darwinian struggle the agent of beauty and a source of wonder, and by then John Burroughs (Muir's contemporary and in his day as popular) set out, in essays that filled several volumes of his collected works, the secular theology he called his "Gospel of Nature."[59] Burroughs believed science swept away our old beliefs, for it "begets a habit of mind in which these artificial notions [belief in the "magic of Christ's blood and all the pagan notions of heaven and hell"] cannot live. . . . The study of nature kills all belief in miraculous or supernatural agents not because it proves to us that the things do not exist, but because it fosters a habit of mind that is unfavorable to them, because it puts us in possession of a point of view from which they disappear."[60]

Science did not sweep away Nature, though, which remained our touchstone of reality, and Muir and his disciples established wild nature as the place to find reality. That focus remained. In the 1950s, in books like *The Singing Wilderness,* Sigurd Olson, one of the founders of the Wilderness Society, called people to search in wilderness for an immediate and emotional grasp of intellectual truths about relationship to the world, and the next generation enshrined the wilderness journey as secular pilgrimage. Like supplicants entering a shrine, travelers left daily life and ordinary conveniences, discarding what would come between them and an experience of the world, to encounter nature with their bodies and muscles, and face the ordinary dangers of river rapids, heat, and cold in search of insight.

Environmentalism looked for enlightenment beyond society, but also within it and within ordinary lives as well. Bioregionalism explored the possibilities of life in contact with local nature, life that would heal land and people. Green consumerism made our consumer decisions about food, clothing, cars, and front yard plantings moral choices that could help save, or destroy, the earth.[61] The daily practice of environmentalism looked to what Christian doctrine called the "conversion of life." By paying attention to the consequences of our actions, we made even trivial actions serve a higher goal and remade our lives and characters. Environmentalism spoke of virtue, demanding humility (using that word in its theological sense as knowledge and acknowledgment of one's actual situation) from us as individuals and as a species, justice toward the world by taking into account other species and the land, prudence in the use of natural resources, and temperance in consumption (all these, but particularly the last, a stench in the nostrils of true believers in the American Way of Life).

Environmental appeals to duty and virtue showed how far the movement strayed from the conventional view of humans and nature. As much as the rest of society, it relied on science, depended on human reason, and looked to the here and now rather than a world beyond, but against the view of humans as the sole measure and source of value, it insisted nature had rights and value; against the idea of autonomy, it insisted on our connections to the larger community of the land; and in opposition to the reduction of all values to money, it insisted some things could not be measured on that scale. Even worse, from the common point of view, it believed the world had limits and, therefore, so did individual ambition and progress. Not always consciously, environmentalism appealed to a philosophical tradition that saw wisdom and virtue in our acting in accord with the world's demands. Its roots in the modern culture made environmentalism a threat to the culture's beliefs, for it spoke in familiar terms. If environmentalists had denied human reason as a way to understand the world or focused exclusively on nature's spiritual benefits, opponents could have dismissed them. They could not, and recognized from the start how deep the new movement's challenge to accepted values was. The speakers the National Agricultural Chemicals Association sent out to oppose Rachel Carson challenged her science, but they expressed moral outrage at her belief that we should live with nature, not conquer it, and accused her of undermining the foundations of Western civilization. *The Limits to Growth,* the report on Club of Rome's computer model of the world system, attracted criticism from left and right, for both identified unlimited economic growth with civilization and immediate measures of human welfare with human progress. Reactions to Earth Day followed the same line. Anti-environmentalists accused the movement of wanting to send people back to the caves to live on nuts and berries, or abolish American freedom by destroying property rights, and they associated the movement with the counterculture—and later with New Age spirituality.

Environmentalism accepted the universe on the basis of reason but also looked for a personal relationship with nature, which came to mean wild country, that yielded insight into the universe. In taking up that part of the movement's heritage, we leave currents of thought common to Western civilization for ones more characteristic of, and at times peculiar to, American society. Ralph Waldo Emerson's Transcendentalism, an American brand of Romanticism, formed the base of American nature appreciation, and so of environmentalism's search for insight, but Darwinian biology formed an essential supplement, a darker view with its own cold beauty. Ecology reinforced Romantic intuitions about our connections to the world by showing them in detail, and changed nature from an arena of beautiful things into systems of intricately interconnected processes. By 1940 a few people had sketched out the intellectual path toward the environmental view and had shown, at least in outline, its consequences. People began following that path in large numbers a generation later. They began in the wilderness, which occupies chapter three.

2 / Emerson's Children

As a reform, environmentalism followed conservation's program of managing natural resources by science and held to preservation's goal of saving wild areas for their beauty. As an expression of the human impulse toward religion, it went beyond arranging our use of nature to ask how we were related to it and what need we had for it, warning that in destroying the wild we killed something in the human spirit. Environmentalism spoke to the human need for a place and purpose in the universe and looked for a way of living with nature inspiring enough to guide people over a lifetime and change society's goals as well. Even at these depths, though, it spoke from within American society and Western culture.

Its vision of nature as part of a human life came from a hundred years of nature appreciation, its expression from Romanticism as interpreted by Ralph Waldo Emerson and his disciples. Environmentalism needed to move beyond the individual spiritual quest to find ultimate reality in nature to form social values that would allow us all to live in harmony with nature—a task for religion. Its vision developed through the generations, as each used current science to find beauty and reality in nature. Thoreau relied on natural history, Muir on natural history and geology, later generations on evolution and ecology. The last two were crucial. Evolution showed nature around us as the visible sign of the invisible forces of life, holy because it was the heritage of the ages. Ecology showed our continuing and immediate ties to life, and together these sciences held out the present as the promise of the future, for what would come depended on the survival of what now lived. Environmentalism also used ecology to fashion a moral yardstick for our relations with the land—most famously in Aldo Leopold's land ethic. "A thing is right," Leopold said, "when it tends to preserve the integrity, stability, and beauty of the biotic community. It is wrong when it tends otherwise."[1]

ENDLESS FASCINATION

Environmentalism built on science, but it also appealed to humans' fascination with nature and their desire to understand, and it satisfied these needs in ways the established view did not. That picture, with its vision of triumphant humanity exercising its God-like powers of reason to shape matter, appealed to the intellect and the passion for power and control. The environmental one dealt with home and place and purpose and described a way to take part in the great work of preserving and rebuilding the forces of life. More important, it spoke to the immediate pull nature exercised on many people, a pull they had felt before they had any science lessons. Sights most adults passed over without a second glance and other children, who (unfortunately) were not gifted with curiosity about nature, saw as commonplace struck these children deeply and permanently. A clutch of eggs in a robin's nest, a rabbit sleeping in a drainpipe, a pickerel hanging in the current of a small stream fixed nature in their minds as a place of wonders, and the scent of spruce gum (or sagebrush or wolf willow), the taste of chewed grass ends, the feel of granite boulders stayed with them, perhaps to the end of their life.

We now associate deep feeling for nature with wilderness, but for most people feelings began in more settled surroundings and, even among the eloquent defenders of nature, stayed there. Henry David Thoreau grew up and lived in the town of Concord and made his retreat from society a mile or so from the village and a few hundred yards from the railroad tracks. He may have turned his steps westward because there they ran free, but they ran, pretty much, in Concord. John Muir made the Sierra Nevada his home as an adult, but came to nature in boyhood rambles along the shore and in the fields of Scotland and sharpened his sense of nature on the Wisconsin farm of his adolescence. Alfred Russell Wallace, who went on to collect in Malaya and define the bioregional division named for him, Wallace's Line, learned botany while surveying in Britain. He described "the solace and delight of my lonely rambles among the moors and mountains" that came with "my first introduction to the variety, the beauty, and the mystery of nature as manifested in the vegetable kingdom."[2] Young Charles Darwin avidly collected beetles. Modern naturalists nourished their passion for nature in the suburbs, along canals, and in visits to parks and playgrounds.[3]

That sort of immediate, personal involvement marked classic nature writing, which lay at the intersection of landscape, memory, observation, and natural history. Gilbert White's *Natural History of Selbourne* (1787), the genre's founding text, had a quite literally parochial setting (the parish of Selbourne), but took in a wider universe, connecting creatures to the land through the new knowledge of natural history and describing the memories of life and events that tied people to the land. Thoreau appointed himself an inspector of snowstorms and surveyor of Concord's fields and pastures, observed Walden Pond winter and summer, measured its depths, listened to its ice creak, and drew his water from it. He walked the town in all seasons and weather, ate frost-chilled apples from abandoned trees, and wrote of plants and animals on every scale from leaves to whole stands of trees.

John Muir took a larger area, the Sierra Nevada, but studied it as Thoreau did Concord, scouting for glaciers and surveying the woods, not measuring board feet but the effect of climate, altitude, and weather on the forest's composition. Like Thoreau taking the best of his neighbors' farms while the neighbors believed he had carried off a few frostbitten wild apples, he took the mountains at their price, at any price—took (like Thoreau) all but the deed, and made *The Mountains of California* a real estate brochure for a spiritual market, a study in the geography and advantages of Eden. Chapters treated various aspects of the property (glaciers, snow, passes, glacier lakes), pointed up highlights (a windstorm, the Douglas squirrel, the water ouzel), and clinched the sale with a description of the vanished Beulah land of California's bee pastures.[4]

Twentieth-century writers also found landscapes they identified with and lived in. In the 1920s, Henry Beston had his Outermost House on Cape

Cod; during the Depression and World War II, Aldo Leopold had his farm in Wisconsin's sand counties; and, a generation later, Annie Dillard had her hermitage along Tinker Creek, just outside Roanoke, Virginia. Recent writers have deliberately turned to what does not seem to be "nature." In *The Meadowlands* (1998), Robert Sullivan uses nature writing's conventions to show beauty and variety in the junk heaps and waste dumps of the great New Jersey swamp just east of New York City.

As well as naturalists' concern with particular places, environmentalism inherited views of nature from every stage of scientific study. Gilbert White wrote at the beginning of natural history, a generation after Linneaus had established the basis of scientific taxonomy, Thoreau under the spell of Humboldt's biogeography and Louis Agassiz's glacial theory and while Darwin was writing *The Origin of Species*. Evolution shaped Robinson Jeffers's poetry, which influenced some of the Sierra Club's leaders and the National Park Service's policies on predatory mammals and introduced species. Aldo Leopold took another step, making ecology the basis of a moral treatment of the land. Environmentalism emerged as a visible social movement and political cause in the 1960s, but its heritage stretched back into American history.

AMERICANS AND NATURE

In English the word "nature" has many meanings, and Americans use them all. They see "nature" as the essential core that makes something what it is, as in the phrase "human nature," and they also make it the label for everything outside humans and their society, "nature" as opposed to culture. That divided gaze went back to the first settlements, when the colonists pictured the land beyond their field as a place of wonder or horror, a storehouse of minerals and furs, the material for fertile fields and rich pastures or the haunt of disease, wild beasts, and wild men. They looked at the Indians in the same fashion, as Nature's noblemen or demons from the pit of Hell. Taking "nature" to mean the system of the world, they justified secession from Great Britain by an appeal to "the laws of Nature and of Nature's God," and, after independence, called in Nature, as the country beyond settlement, to make up for the deficiencies of culture—the lack of high art and monuments. America possessed greater rivers and loftier mountains than Europe and offered farms for all.

Catherine Albanese described early-nineteenth-century Americans' fascination with "Nature" in all its forms—physical world, abstract principle, ground of reality, emblem of the higher spirit—as the American nature religion. She viewed this conjunction of individual searches for the sacred as part of Americans' search for harmony, mastery, and freedom, "one more sign that, in a 'secular' society, the search for the sacred refuses to go away."[5] This faith did not have any special place for wild nature—indeed, it was not particularly concerned with "nature" in that sense. Looking at how people related to the world rather than functioned in it, the American nature religion was often "about place and mastery *in society*."[6]

People at the time, she admitted, did not see the elements she put together as a single thing, and her description was less a "contemporary social construction of past and present American religion" than a useful way for the historian to see links, connections, logic, and the power of ideas. The enthusiasm for "nature" suffused the movement for a pure Christianity that split so many Protestant churches in the early part of the century, and it underlay the campaigns for health and medical reforms—homeopathy, naturopathy, and the other competitors of what we would now consider conventional medicine. Nature religion permeated mesmerism (hypnotism) in the early part of the nineteenth century. Mesmerism was, indeed, "one of the controlling metaphors of American nature religion."[7]

Some Americans, though, turned to nature in the ordinary sense of plants and animals, woods, fields, and forests, seeking there the opening to an extraordinary world. One, Ralph Waldo Emerson, established nature (which his disciples came to see as wild nature) in American culture as part of a full human life, the path to self-knowledge, and our way to understand the universe. He told a generation embarked on an experiment in democratic society in a new nation to turn from Europe to their own land and to abandon formal systems in favor of their own experience. He hymned, or seemed to hymn, the individual, the American Adam in the Garden of the World, to say that history and tradition (of which America had little) were of no importance, that nature (of which America had an abundance) was all we needed. With compelling metaphors and striking aphorisms, he described nature as refuge, instructor, and source of wisdom. A later generation made Henry David Thoreau and John Muir the first nature saints, but they were Emerson's disciples— Thoreau his protégé, and Muir developed intellectually in a circle influenced by Emerson.

Emerson regarded Nature as at once the radically Other and the gateway to ourselves. Outside society, unaffected by humanity's tricks and shams, nature did not deceive. Uniquely real, it transforms us. "Standing on the bare ground—my head bathed by the blithe air and uplifted into the infinite space—all mean egotism vanishes. I become a transparent eyeball; I am nothing; I see all; the currents of the Universal Being circulate through me; I am part or parcel of God." According to Emerson, nature stands outside us, but it is also part of us and calls us. "The greatest delight which the fields and woods minister is the suggestion of an occult relation between man and the vegetable. I am not alone and unacknowledged. They nod to me, and I to them." Our ties to Nature were more than personal or physical—they were metaphysical. Two elements divided the world between them, Nature and Soul, and there is "a radical correspondence between visible things and human thoughts. . . . Every natural fact is a symbol of some spiritual fact. . . . The whole of nature is a metaphor of the human mind. The laws of moral nature answer to those of matter as face to face in a glass."[8]

By their works, but more by the examples of their lives, Thoreau and Muir established a tradition of nature appreciation on a modified Emersonian philosophy. They had little difficulty picking and choosing, for Emerson was not a systematic philosopher or an exact writer, and his essays did not so much present ideas or develop arguments as set out quotable musings on declared truths in vivid phrases. But while his disciples changed some of Emerson's concepts they held to others—especially the correspondence between humans and the world: Soul and Nature. For Emerson, going to nature meant going home; we labor in society, but long for Nature (now suburbia forms the vale of tears and wilderness the New Jerusalem, but the terms are recognizably Emersonian). Emerson's disciples accepted as well his distinctly Protestant and American stress on individual, untutored contact with an intelligible world. Emerson sent people to nature in the same spirit and with the same confidence that Protestant ministers directed people to the Bible. (This is not a casual example. The American nature movement began with failed Protestants, and they still have a strong influence.[9]) As any man of faith could read the Scriptures, so any earnest searcher could understand nature. We needed no guide, no priest, no tradition; our instincts and good hearts would serve.

Believing, with Emerson, that books were for the scholar's idle hours, his disciples emphasized experience—Thoreau in Concord, Muir in the Sierras. They found, however, a different "nature" than Emerson's. Emerson saw the universe as "composed of Nature and Soul"—Soul being the primary component. Nature—everything material—was of no particular significance. It would, Emerson thought, be as useful whether it had "a substantial existence without, or [was] only in the apocalypse of the mind." It was the touch of humanity that counted. "All the facts in natural history, taken by themselves, have no value, but are barren, like a single sex. But marry it to human history, and it is full of life. Whole floras, all Linnaeus', and Buffon's volumes, are dry catalogs of facts; but the most trivial of these facts . . . applied to the illustration of fact in intellectual philosophy, or in any way associated to human nature, affects us in the most lively and agreeable manner." His disciples reversed that emphasis, making Soul secondary and Nature primary. They saw Nature as the biological and physical world beyond human society (beginning the emphasis that would lead to the wilderness movement), and found interest in nature not because it was married to human history but because it was beyond humans. The touch of Nature, they felt, gave life to Soul—a belief that still shapes the understanding of millions who have never read Emerson.[10]

The rise of Emersonian nature in American culture began with America's move to the city. As wilderness and wild country dwindled and Americans faced an industrial economy, an urban future, and cities filled with unsuitable immigrants, "old-stock" Americans found in Thoreau a prophet and an exemplar of spiritual quest in a realm identified with their group—nature. When Thoreau died in 1862, he had been a minor figure on the New England literary and intellectual scene—much less important than his friend and mentor, Emerson. By the end of the century, changing times brought pilgrims to Walden Pond and established *Walden* as a sacred text. Thoreau's importance grew with the continued loss of wild land and wild nature.[11] A replica of his shack now stands by Walden Pond's parking lot, and at its site a board engraved with the words: "I went to the woods because I wished to live deliberately, to front only the essential facts of life, and see if I could not learn what it had to teach, and not, when I came to die, discover that I had not lived."[12] On a visit in the summer of 2000, I tiptoed past a pilgrim brooding at the shrine.

People seized on Thoreau's life as an example, but especially his concern for "wildness," which he saw as the ground and center of life itself. "In Wildness is the Preservation of the World," he said; nature lovers, then environmentalists, cited that as Scripture and put it on posters (usually with an Eliot Porter photograph) while the Sierra Club used it as the title for one of its lush picture books.[13] Thoreau's "wildness" was not "wilderness," though people confuse the two often enough that, at times, they misquote Thoreau as saying, "In Wilderness Is the Preservation of the World." Not so, but misquotation is not misinterpretation.[14] The slip only emphasized or overemphasized Thoreau's belief that we require contact with the life of the world if we are to live fully. The modern wilderness movement built on that belief. Focusing on places outside society, it emphasized direct experience rather than interpretations of what we see.[15]

Generations of English professors, helped by historians, have emphasized Transcendentalism's intellectual qualities, but Emerson and his circle were less concerned with forming a view of the world than responding to it. They did not look for the quiet satisfaction of a logical understanding but for a glimpse of the world's foundations. Robert Richardson caught that passion in the subtitle of his biography *Emerson: The Mind on Fire,* which began with the twenty-eight-year-old Emerson's visit to the grave of his wife (dead more than a year), where he opened not just the vault, but the coffin. When Emerson called on people to form an original relation to the universe, he wanted, as Richardson put it, "direct, personal, unmediated experience."[16]

Thoreau's famous passage about Mt. Ktaadn showed the same passion. The mountainside was "made out of Chaos and Old Night. Here was no

man's garden, but the unhandseled globe . . . the fresh and natural surface of the planet Earth. . . . Man was not to be associated with it. It was
Matter, vast, terrific—not his Mother Earth that we have heard of, not
for him to tread on, or be buried in—no, it were being too familiar even
to let his bones lie there . . . to be inhabited by men nearer of kin to the
rocks and wild animals than we. . . . Talk of mysteries! Think of our life
in nature—daily to be shown matter, to come into contact with it—rocks,
trees, wind on our cheeks! the *solid* earth! the *actual world!* the *common*
sense! *Contact! Contact! Who* are we? *Where* are we?"[17] That was the mysterium tremendum, the awe that shook the soul faced with what was utterly
beyond humanity. Thoreau inverted Emerson's categories, making Soul
the ordinary, Nature the mysterious, but he spoke to the Transcenden-
talists' central quest for reality, what lay behind the world's familiar face
and ordinary routines.

Transcendentalism also incorporated science, for natural history,
unlike the modern biological sciences, looked beyond facts for meaning.
Thoreau followed a spiritual path not by standing on hilltops as a transparent eyeball, with the currents of Universal Being circulating through
him, but by working in natural history. He listened to Louis Agassiz's theories about an ice age, read Darwin's *Origin of Species,* and during the last
decade of his life undertook an ambitious program of natural history, a
subject that then covered the entire natural world from soil to clouds.[18]
He followed the great German naturalist Alexander von Humboldt, who
believed natural history's data led not only to understanding but to wisdom, a program Humboldt most clearly expressed in his *Cosmos* but that
appeared in all his works.[19] Once we knew the butterfly and the mountain, Humboldt believed, we would know the ties between them and their
connections to us. Thoreau, believing with Humboldt that "man and
nature were on every level dependent on and expressive of each other,
and the 'facts' of nature were energetic co-productions of the human mind
operating with and within the field of natural objects," made charts and
tables of when plants flowered and fruited in Concord, when ice formed
on the ponds and when it broke up, leaving at his death an unequaled
botanical and meteorological calendar of his town.[20]

Thirty years later John Muir used science to understand nature in the wilder country of the Sierra Nevada, and his joyous reports made him nature's prophet to an industrializing nation. Like Emerson and Thoreau he began in conventional Protestantism, his father's Campbellite faith, ever harsher than New England Congregationalism, and ended in nature. Nature's pull came early. Muir recalled as a child in Scotland wandering "in the fields to hear the birds sing, and along the seashore to gaze and wonder at the shells and seaweeds, eels, and crabs in the pools."[21] Around his father's Wisconsin farm (the family emigrated when he was eleven), Muir received a "baptism in Nature's warm heart . . . every wild lesson a love lesson, not whipped but charmed into us. [The whipping had helped him memorize the New Testament and much of the Old before he was twelve.] Oh that glorious Wisconsin wilderness!"[22]

At the University of Wisconsin he came across Emerson's ideas and science—particularly botany, which he made his first tool for exploring nature. He left without a degree, botanized, drifted, and worked as a mechanic until a machine shop accident in Indianapolis left him blind for some months. Abandoning the shop, he set off to walk to the Gulf of Mexico, planning—like an entire generation before him—to follow Humboldt's example and explore South America (Humboldt's *Travels* had also fired the youthful ambitions of Charles Darwin and Alfred Russell Wallace). Illness and ship schedules diverted him to California, where he made his way to the Sierra Nevada and found Yosemite and his life's passion: the great mountain range running the length of the state. He worked and wandered, discovered most of the range's glaciers, discovered Yosemite Valley's glacial origin, and made a reputation as a field geologist. He began writing in the 1870s, and two decades later the combination of an urban population looking for nature and a publisher looking for money (Robert Underwood Johnson) brought him fame.

In prose steeped in Old Testament rhetoric, informed by Romanticism and science, Muir preached the Emersonian gospel of Nature as ultimate reality, refuge from society, and place of pilgrimage, his life giving authority to his call to come to the mountains, lay down your burdens, and be refreshed, and rejoice in the beauties and glories of creation. He spoke of the mountains as God's Temples and the Sierra meadows as a Garden of Eden and condemned the flooding of the Hetch-Hetchy Valley to provide drinking water for San Francisco as desecration. Unlike the prophets, though, Muir comforted rather than challenged. He rejoiced that "thousands of tired, nerve-shaken, over-civilized people are beginning to find out that going to the mountains is going home . . . wilderness is a necessity, and mountain parks and reservations are . . . fountains of life."[23]

Rather than calling people to a new life in nature, though, he told them to make nature part of their lives. Thoreau stood in nature and criticized society, demanding that we see the conventional pieties as hindrances to real life. Muir asked that we put aside conventional pieties for the moment to ramble in the wilderness. Thoreau insisted there were real evils in the way we lived. Muir eloquently described particular problems but damned society only when it destroyed nature's temples, and even then said nothing about the forces that made for destruction. He railed against grazing in the high mountains, calling sheep "hoofed locusts," but not against the institutional arrangements and social values that brought the animals to the high meadows.

Muir spoke in conventional and conventionally pious terms, mixing the vocabularies of natural theology with terms from geology and biogeography. He shifted the focus of awe and adoration from Nature's God to Nature so gently that few noticed, none were disturbed, and scholars still debate when or whether he abandoned the Christianity of his youth.[24] He described the range as the product of vast internal forces; rain and snow shaped them. Rain, sun, the seasons, and the altitude set the trees, grass, and flowers each in its own place.[25] No wild animals threatened the pilgrim, only the calculable dangers of storms, snow, and rockfall; all was alive and all worked toward good. "One fancies a heart like our own must be beating in every crystal and cell, and we feel like stopping to speak to the plants and animals as friendly fellow mountaineers."[26] Only the bloodless competition of plants for light and space suggested to him anything that might be called the "struggle for existence," and Muir did not even hint at the things that led Darwin to exclaim: "What a book a devil's chaplain might write on the clumsy, wasteful, blundering, low and horribly cruel works of nature."[27]

Of nature's destructive power Muir would not hear; Death comes gently and "everything in Nature called destructive must be creation—a change from beauty to beauty."[28] Like the original Eden, this one seemed barred to humans. Muir called people to nature, but he did not regard them as worthy of it. Those who worked among the mountains seemed oblivious to their wonders, and visitors to Yosemite seemed "wholly unconscious of anything going on about them, while the sublime rocks were trembling with the tones of the mighty chanting congregation of waters."[29] Returning bedraggled from a day spent in a storm, he found people pitied him as if he were a castaway, but he pitied them "for being dry and defrauded of all the glory that Nature had spread round them that day."[30] Beyond blindness to beauty, though, he saw some essential flaw in humankind. Even the Indians, so often stereotyped as a people attuned to nature, seemed unworthy of the mountain landscape. Meeting a group of them, he wondered at the "strangely dirty and irregular life these dark-eyed, dark-haired, half-happy savages live in this clean wilderness."[31] Even the necessities of life offended. "Man seems to be the only animal whose food soils him."[32]

By the time he died in 1914, Muir was an icon in the pantheon of American nature saints, and as the years passed his reputation grew. His writings became part of American nature literature, his life in the mountains an inspiration, and his Sierra Club a leading environmental organization. He deserved his reputation for, more than anyone else, he made contact with wild nature a part of modern America and the preservation of wild county an acceptable and accepted land use. He achieved this, though, by using the culture against itself, and the ironies of his crusade suggested some of the problems environmentalism would encounter half a century later. He preached the gospel of nature in magazines produced on high-speed presses, shipped across the nation on railroads, and financed by advertisements for consumer goods. The money and political influence that made the Sierra Club a power in the land came from San Francisco's commercial elite, the men who directed the companies that cut down trees, dug mines, and built railroads. His hymns to the mountains' glories brought a flood of visitors who shattered the peace they came for.

In the early twentieth century these contradictions did not loom large, for people could believe that progress and preservation need not clash, that incidents like San Francisco's damning the Hetch-Hetchy Valley in Yosemite for a reservoir would be rare and regrettable exceptions. Development could flow around the national parks, leaving them as islands of nature in the developed countryside. Fifty years later, as ecology showed the ties between progress and environmental degradation and events made it clear that park boundaries alone could not protect the wild, environmentalism grappled with problems Muir's generation had finessed and with the limits of its solutions.

Environmentalism faced another conflict Muir's generation could ignore, between a spirituality of nature and a science that threatened to strip all spiritual values from it. Thoreau, writing in the heyday of natural history and before Darwin, found that science fit his needs, but Muir had to choose among the developing disciplines. He used biogeography (one of Humboldt's great contributions) and geology but avoided evolution. Nature lovers, though, could still enthusiastically embrace both. Muir's contemporary, John Burroughs, managed that, and his essays on what he called his gospel of nature provided an explicitly religious view that millions read. Burroughs's reputation has dwindled, but in the early twentieth century "John o' the Birds" had a following to rival "John o' the Mountains." Muir accompanied Teddy Roosevelt through Yosemite; Burroughs went camping and tramping with the president in Yellowstone. Muir made the Sierra a place of pilgrimage, but after Burroughs's death in 1921, his cabin, Slabsides, became a shrine and his writings appeared in a multivolume standard edition.

Unlike Muir, Burroughs developed his ideas from a broad concern with the culture. He had been one of Walt Whitman's early champions, was the last of Emerson's serious interpreters, and wrote extensively on literature and popular philosophical questions. He also embraced science, and in essays written over almost twenty years and filling two volumes of his collected works (*Light of Day* and *Accepting the Universe*), Burroughs set out his views.[33] The march of Reason and Progress, he declared, had overtaken our ideas; we should turn from outmoded theologies (belief in the "magic of Christ's blood and all the pagan notions of heaven and hell") to the common light of day, the only sound base for a faith and a life in

the modern world. "Science begets a habit of mind in which these artificial notions cannot live. . . . The study of nature kills all belief in miraculous or supernatural agents not because it proves to us that the things
do not exist, but because it fosters a habit of mind that is unfavorable to
them, because it puts us in possession of a point of view from which they
disappear."[34]

That point of view did not make nature disappear. On the contrary,
"[n]ature love as Emerson knew it, and as Wordsworth knew it, and as
any of the choice spirits of our time has known it, has distinctly a religious value," a value Burroughs found in Emerson.[35] According to Burroughs, nature was reality: "In intercourse with Nature you are dealing
with things at first hand, and you get a rule, a standard, that serves you
through life. You are dealing with primal sanities, primal honesties, primal attraction, you are touching at least the hem of the garment with which
the infinite is clothed."[36]

Like Emerson, Burroughs called us to face the universe, and for that
we would need "encouragement in our attitude of heroic courage and faith
toward an impersonal universe; we need to have our petty anthropo-
morphic view of things shaken up and hung out in the wind to air."[37] The
universe is stern—here Burroughs returned to the iron discipline of his
father's Calvinist faith—"Nature is not benevolent; Nature is just, . . .
makes no exceptions, never tempers her decrees with mercy, or winks
at any infringement of her laws. And in the end is not this best? Could
the universe be run as a charity or a benevolent institution . . . ?"[38] Science did not destroy the Emersonian faith; indeed, science reinforced it.
"Viewed as a whole," Burroughs said, "the universe is all good. . . . This
is not the language of the heart or of the emotions . . . it is the language
of serene, impartial reason."[39]

Burroughs mixed science-based rationalism and Romantic faith in
nature for a generation coming to see science as a source of authoritative
knowledge about the world but still looking for purpose in the universe.
He made a rational but comforting faith that served many as a religion
or a substitute for one. *The Gospel of Nature* is, in fact, still in print, in a
booklet about the size and shape of the ones in the pamphlet rack in the
back of the church. It has a genuinely religious view, for it confronts ulti-

mate questions, but it is hardly complete. Burroughs suggested a spiritual path, but said little about the truths that should guide the pilgrim seeking to live in tune with the universe. He provided less an alternative to accepted views about how to live than a supplement to them, in which nature made up for the deficiencies of society. He said little about the society and less about society's relation to nature, placing them, as Emerson had, in separate spheres.

SACRED TIME, SACRED SPACE, SACRED LAND

Environmentalism inherited Emerson's ideas and in the twentieth century developed the concept of a sacred nature to which humans were bound.

Even as Burroughs preached the gospel of nature, its foundations shifted. Emerson, his polestar, dwindled from cultural force to literary influence and classroom assignment, and natural history dissolved into the specialized disciplines of the modern earth and biological sciences. Burroughs described the rural landscapes around his Catskill farm, but Americans were leaving the farm, and when they looked from the city to nature they turned toward wild country (the decline of Burroughs's reputation and the continued rise of Muir's owed much to this change). But while the foundations shifted, they remained strong. Emerson, albeit Emerson as interpreted by Thoreau and Muir, remained Americans' guide, and science continued to be their authoritative source of knowledge about the world.

The great changes in American devotion to nature came from using evolution to show its sacredness and ecology to guide our treatment of it. These might seem impossible since science insisted on material causes and resolutely ignored moral questions and ultimate ends, describing how the world worked, not what we should do to it. It might help people take delight in the world's beauty and complexity, but it would seem useless in the search for meaning. Since the seventeenth century, though, people had enlisted the facts of science in the search for meaning in the world, and while Darwinian evolution made simple and sunny versions of natural theology impossible, it did not bar more sophisticated, if darker, explanations.[40] Indeed, it encouraged them, for there was, as Darwin famously

said, a certain "grandeur in this view of life"—a world in which continuing struggle produced from simple forms all the beautiful and varied
things we saw.

Nature writers quickly seized on this perspective. In the 1890s Ernest
Thompson Seton and Charles G. D. Roberts made their reputations with
animal stories describing how the round of struggle and death produced
nature's beauty. Seton worked by tempering the Darwinian perspective
and fitting it to conventional morality. He acknowledged the fact of
inevitable death in nature, but preserved a happy ending by cutting his
stories off before that point. He argued that species that followed the Ten
Commandments were more successful in the struggle for life than those
who did not—nature favored monogamous and faithful species over pro-
miscuous ones—and used his stories to teach moral lessons. Young animals that observed their mothers' command to keep still, for example,
lived, because predators did not find them. Those who failed to heed their
mothers' warnings died. He appealed to accepted values by putting the
struggle for freedom at the center of many of his stories. Roberts faced
more directly the horrors and cruelties so evident to naturalists. Some of
his stories were simply strings of death, with the victor of one struggle
becoming victim in the next and death a common fate that even mother
love could not stave off.[41] Roberts's work, possibly for those reasons, was
less popular than Seton's.

While the nature writers focused on the struggle of life, other defenders of nature used evolution's changes to show how nature was worthy
of our awe. The plants and animals and the land itself were the heritage of
the ages, the product of forces going back to the origins of life on earth
that would shape it until the sun died, the visible sign of the invisible grace
of life's continuing work. The poet Robinson Jeffers was one of the first
to adopt that point of view, and his poetry influenced a number of significant environmental thinkers and activists. He was popular in the 1920s
among a group that included Joseph Campbell and John Steinbeck, fell
out of fashion during the Depression, alienated the reading public by stead-
fast isolationism during World War II, and, by the time he died in 1962,
had passed from the critics' view.[42]

A few found his work a revelation. William Everson, also a poet, wrote of his discovery of Jeffers as a "religious conversion, intellectual awakening, and artistic birth." Loren Eiseley and Gary Snyder acknowledged his impact, and Ansel Adams said he was trying to do with photographs what Jeffers had done with words—show the beauty of the world.[43] In 1965 the Sierra Club published a book, *Not Man Apart,* with the subtitle *Photographs of the Big Sur Coast, Lines from Robinson Jeffers,* and David Brower, who wrote the introduction to that book, took Jeffer's phrase "not man apart" for the Friends of the Earth's newsletter.[44]

The son of a Protestant theologian, Jeffers received a classical and philosophical education, which he supplemented with studies in medicine and forestry before settling on the California coast at Carmel in 1914. He used the common conventions of nature writing—the rhythms of nature, from daily through geological time, and the rocks, waves, and wildlife, seen closely in a particular area—but Everson made the case, in *The Excesses of God: Robinson Jeffers as a Religious Figure,* that Jeffers's search was ultimately religious: "To participate in reality by virtue of its divine character is the central theme of his testament. Transcendence as the mode of ultimate realization is the basic program of all he advocates."[45] A vision of "'God' (the quotation marks are necessary, given Jeffers' deep alienation from conventional views and established religion)" dominated his work. Jeffers felt acutely his own "nothingness, a nullity of self realized in contrast to that which is infinitely beyond it, to a being inexpressibly supreme above all creatures." His stance was not a philosophical or considered position, but "a disposition primarily and consistently of the heart and only secondarily and inconsistently of the mind."[46]

Jeffers himself said that he believed "the universe is one being," all its parts connected, communicating and "influencing each other, therefore parts of one organic whole. (This is physics, I believe, as well as religion) . . . . This whole is in all its parts so beautiful, and is felt by me to be so intensely in earnest, that I am compelled to love it, and to think of it as divine. It seems to me that this whole alone is worthy of the deeper sort of love; and that there is peace, freedom, I might say a kind of salvation, in turning one's affections outward toward this one God, rather than inward on oneself, or on humanity, or on human imagination and abstractions—the world of spirits."[47]

Jeffers's view stood out for its emphasis on the pain, cruelty, and death involved in life, its close focus on a phenomenon naturalists saw and shuddered at. As Annie Dillard put it in *Pilgrim at Tinker Creek:* "Fish gotta swim and bird gotta fly; insects, it seems, gotta do one horrible thing after another."[48] The wasps that paralyzed their prey and left it in a burrow to be eaten alive by their larvae took first place on most people's lists of nature's horrors. "They would justify," Loren Eiseley wrote, "Darwin's well-known remark about the horribly cruel works of nature, or even Emerson's observation that there is a crack in everything that God has made."[49] Naturalists, though, averted their gaze and passed on. Muir linked destruction in nature to the production of more beauty, and nature writers turned to the comfort of nature's recurring rhythms. In *The Outermost House* (1928), an interwar nature classic that was one of Rachel Carson's favorites, Henry Beston spoke lyrically of the "great rhythms of nature" that showed a "world whose greater manifestations remain above and beyond the violences of men," and Donald Culross Peattie's *Prairie Grove* (1940) had the same theme.[50]

In sharp contrast to that convention, Jeffers refused to pass by or dismiss pain and cruelty as minor elements. He saw them instead as central to life. "What but the wolf's tooth whittled so fine / The fleet limbs of the antelope? / What but fear winged the birds, and hunger / Jeweled with such eyes the great goshawk's head?" He believed that to be was to suffer, and insight came only "through experiencing anguish and identifying with the process of suffering."[51] He asked us to "praise life, it deserves praise, but the praise of life / That forgets the pain is a pebble / Rattled in a dry gourd" ("Praise Life").

Even more than Muir, Jeffers measured people against nature and found them wanting. Muir's distaste was occasional, and he called us to join in Nature's happiness. Jeffers seemed to regard humanity as a blight on nature, its works smudges on nature's vast beauty, and to look forward to the day the tide would cleanse the land and the rubble of human life would return to dust.[52] Critics often accused him of hating people (that he called his philosophy "inhumanism" did not help his case), but that was too simple. Jeffers celebrated life and work that respected nature. It was in the Carmel country, he said, that he first saw "people living [a] life purged of ephemeral accretions. . . . Here was contemporary life that was also permanent life, and not shut off from the modern world but conscious of it and related to it, capable of expressing its spirit but unencumbered by the mass of poetically irrelevant details and complexities that make a civilization."⁵³ Of the boats coming into Monterey Harbor he wrote, "A flight of pelicans / Is nothing lovelier to look at; / The flight of the planets is nothing nobler; all the arts lose virtue / Against the essential reality / Of creatures going about their business among the equally / Earnest elements of nature" ("Boats in a Fog"). William Everson argued that Jeffers's apparent distance from humans and seeming contempt stemmed from his prophetic stand. Mystic as well as prophet, and as such calling us to repentance, drunk or ecstatic with the beauty of the world around him, Jeffers would stop "at nothing to confront the race to itself so that it might turn from itself and love God." His was "the intolerable loneliness of the religious spirit utterly isolated by the stupendous truth of its vision."⁵⁴

Few fully embraced Jeffers's dark view, but many believed with him in a nature that was sacred because it expressed the forces of life, and they used that view to argue for changes in public policy. In 1933 George Wright, head of the National Park Service's new Division of Wildlife, called for the parks to preserve every species, for each was the "embodied story of natural forces which have been operative for millions of years and is therefore a priceless creation, a living embodiment of the past."⁵⁵ This set a new mission for the National Park Service and for the country. When the agency had noticed animals, it regarded them as mobile scenery, a spectacle for the tourists. It protected the large herbivores like deer, elk, and buffalo, and the "cute" bears tourists looked for along the road. It ignored the rest—except for predatory mammals like coyotes, wolves, and mountain lions, which it poisoned or shot on sight. Even nature's staunch defenders drew the same line. The Audubon Society banned the "bad," or bird-eating, hawks from its refuges at the muzzle of a shotgun and only held back from condemning those species for fear that too many people would kill the "good," or rodent-eating, hawks as well as the "bad," bird-eating ones.⁵⁶ Calling for these species to be preserved—and Wright did mean that—involved a new vision of nature.

Evolution made nature sacred by connecting what we saw to the birth of life on earth. Ecology made nature a matter of moral concern by showing how our daily lives were part of the "web of life." Environmentalists usually (and rightly) credit Aldo Leopold with making our connections to the land clear and see this land ethic as the clearest expression of our moral duties toward nature. Radical as the land ethic was, though, it drew on a line of thought that began with George Perkins Marsh's *Man and Nature* (1864). That book argued that humans constituted a worldwide geological and biological force and warned that we were destroying the land on which we depended for life.

Nature is not, Marsh said, inexhaustible, and humans have had more than local and temporary effects. Ancient civilizations had reshaped the lands of the Mediterranean, and we were making more and more dramatic changes. We are "even now, breaking up the floor and wainscoting and doors and window frames of our dwelling, for fuel to warm our bodies and seethe our pottage, and the world cannot afford to wait till the slow and sure progress of exact science has taught it a better economy." We had to change our ways. Marsh appealed to self-interest—the survival of civilization—but rested his case on morals. Accepting, even celebrating, human dominion over the earth and believing that God had placed us here and put all things under our feet, he nevertheless saw limits to our dominion. "Man has too long forgotten that the earth was given to him for usufruct alone, not for consumption, still less for profligate waste."[57]

Progressive conservation as embodied in Gifford Pinchot's national forest policies measured morality by economic efficiency, but older standards remained, and they were central to the arguments against growth that developed during the Depression and continued into the boom years after World War II. In *Deserts on the March* (1935), Paul Sears, speaking against the backdrop of the dust bowl, condemned American civilization's profligate ways. "The lustful march of the white race across the virgin continent, strewn with ruined forests, polluted streams, gullied fields, strained by the breaking of treaties and titanic greed, can no longer be disguised behind the camouflage which we call civilization." We must find new ways of treating the land, policies with something better than the "perspective of the newborn," with a "sense of continuity and proportion."[58]

Books like William Vogt's *Road To Survival* and Fairfield Osborn's *Our Plundered Planet,* both published in 1948, argued for public policy in moral language. Humans, Vogt warned, had "backed themselves into an ecological trap. . . . [and were living] on promissory notes. Now, all over the world, the notes are falling due. . . . Unless . . . man readjusts his way of living, in its fullest sense, to the imperatives imposed by the *limited* resources of his environment . . . we may as well give up all hope of continuing civilized life. Like Gadarene swine, we shall rush down a war-torn slope to a barbarian existence in the blackened rubble."[59]

Each generation had more of that crushing modern argument, statistics, and more ecological information tying human actions to changes on the land. Aldo Leopold did not use figures but did use ecology's theories as the foundation for a morality of human action on the land. He saw the links among humans, other animals, plants, and the soil itself binding us into a community, a whole in which each species has a function, place, and purpose. Because humans are part of the community of the land and so affect others, he argued, they have the obligation to protect and preserve it. That made the health of the land, the moral measure of action. "A thing is right when it tends to preserve the integrity, stability, and beauty of the biotic community. It is wrong when it tends otherwise."[60]

This, Leopold's clearest statement of his land ethic, connected fact and value—the world science showed with what might and ought to be—and while philosophers did not find Leopold's arguments entirely satisfactory that hardly mattered. People lived less by philosophical systems than by applying those systems to their lives—usually in ways that horrified philosophers—and they found the land ethic a useful framework to think with and a good guide to what was important. It was not always apparent what would "preserve the integrity, stability, and beauty of the biotic community," but the problems of deciding what we should do were in principle no different from those encountered in using the Golden Rule to guide our relations with other people.

Leopold did not so much apply morals to our treatment of the land as change the basis of the morality people used. He drew on the existing tradition of stewardship, most clearly from Liberty Hyde Bailey, a horticulturist at Cornell University best known for his advocacy of nature education and the reform of rural life. In 1915, he copied quotations from Bailey's *Holy Earth* into a notebook, and in his book *Game Management* (1933) quoted Bailey. When he came to write his mature essays in the late 1930s, he incorporated Bailey's argument without the Christian concepts and language.[61] Bailey argued that the earth was holy and therefore we had to abandon efforts to make a living from it while leaving the problems we caused to the next generation. We had to "deal with [the land] devotedly and with care that we do not despoil it, and mindful of our relations to all beings that live on it." God gave us dominion over nature, but our situation was one of "obligation and service" to others, future generations, and the earth itself. Public policies were necessary, but our ties to the land had also to be personal and moral. "It would seem that a divine obligation rests on every soul," and "all people, or as many of them as possible, shall have contact with the earth and . . . the earth's righteousness shall be abundantly taught."[62]

Bailey did not argue that everyone should go "back to the land," as the popular slogan and movement of the period had it, but that we should find ways for people to take "personal satisfaction in the earth to which we are born, and the quickened responsibility, the whole relation broadly developed." That meant living in accordance with the world's laws and needs, living as part of the world and in cooperation with it. "A useful contact with the earth places man not as superior to nature but as a superior intelligence working in nature as a conscious and therefore as a responsible part in a plan of evolution, which is a continuing creation."[63] Leopold moved from the Bible to ecology and also—a shift as momentous as the first—extended his concern from fields and pastures to wilderness areas. Both changes grew out of his lifelong learning and continuing involvement in land use on every scale. Looking back, he said that his "earliest impressions of wildlife and its pursuit retain a vivid sharpness of form, color, and atmosphere that half a century of professional wildlife experience has failed to obliterate or to improve on."[64] From hunting and sportsmanship as a child, he moved, as a student at Yale forestry school, to conservation, for the forestry school was then the training ground for Gifford Pinchot's Forest Service and the national conservation crusade he was leading.

Graduating in 1909, Leopold joined the agency and spent fifteen years in the Southwest applying conservation doctrine and seeing its effects on plants, animals, and people. He also read widely and, through his naturalist observations, came in contact with the new science of ecology. By the early 1920s he was looking beyond conservation and even expressing open doubts about the conventional wisdom. Speaking to an Albuquerque civic group in 1923, he offered "A Criticism of the Booster Spirit," pointing to the futility of growth for the sake of growth and calling for something more than dollar values.[65] The next year he left the Southwest for Madison, Wisconsin, and a few years later left the Forest Service to build the new profession of game management. Through the 1930s he learned about farm and woodlot biology and the problems and possibilities of farm living, and more about ecology. On a trip to Germany he saw managed, industrial forests, and in Mexico largely undisturbed ecosystems.

Because the land ethic grew from Leopold's work in applying science, and because he wrote for people without technical or scientific training, he never put his ideas in a philosophically rigorous form. Often, in fact, he used common terms and familiar things to describe new situations. The essay "The Land Ethic," for example, presented ethics as "modes of co-operation," ways to live in an interdependent world, worked out in society by experience in small groups and slowly extended to larger ones, but he began with a concrete example. When Odysseus hung the faithless slave women after his return from Troy no one objected, because ethics did not apply to slaves. We had, the essay went on, extended ethics to slaves, then to people further off, then to society as a whole. The Ten Commandments guided relations among individuals; the Golden Rule tried "to integrate the individual to society; democracy to integrate social organization to the individual."[66]

Now we needed a way of "meeting ecological situations so new or intricate, or involving such deferred reactions, that the path of social expedience is not discernible to the average individual." Self-interest and simple prudence would not serve, and economics was no guide, for "most members of the land community have no economic value." To make an ethical standard that covered the larger community of the plants, animals, and the soil, we had to "quit thinking about decent land-use as solely an economic problem, [to] examine each question in terms of what is ethically and esthetically right, as well as what is economically expedient." The argument was straightforward, the terms familiar, economic concerns acknowledged, but the standard of right a new and radical one.[67]

Leopold recognized the implications of his work. Accepting the land ethic changed "the role of *Homo sapiens* from conqueror of the land-community to plain member and citizen of it," and he warned that "no important change in ethics was ever accomplished without an internal change in our intellectual emphasis, loyalties, affections, and convictions."[68] The land ethic challenged central tenets of the American faith in a new way. The culture had incorporated earlier movements for nature by treating them as individual spiritual experiences pursued within the overall goal of individual fulfillment. It saw Thoreau and Muir as individuals pursuing their dreams, made national parks part of the nation's glories, and even accommodated the wilderness movement by making wilderness the setting for pioneer virtue and outdoor recreation. The land ethic, though, gave the community a central role and made the individual subordinate; and while that did not fit into the idea of individual aspirations it could not be dismissed. It was possible to see Bailey as a reactionary or a charming throwback to our agrarian period, interesting and even praiseworthy, but out of touch with industrial America. Leopold, though, clearly was in touch. Worse, he wrote in clear, accessible prose. It may have been hard to live by his ideas, but it was easy to see what he aimed at.

Leopold did not use explicitly religious language, probably would have been skeptical of an environmental "religion," and surely would have been horrified at the suggestion he was helping to establish one. Yet his work spoke to the religious dimension of life and to ultimate questions and needs, and the public made him, deservedly, an environmental nature saint. Teachers, first in colleges, then in high schools, assigned *A Sand County Almanac*. Academics developed courses around his work, analyzed his ideas, and brought his unpublished papers into print and his uncollected articles together. There is even a volume entitled *The Essential Aldo Leopold*, mercifully not in the same series as *The Essential Rumi/Budda/Christ*.[69] Environmentalists made Leopold's sand county farm hallowed ground. They had reason. The land ethic showed the common founda-tion beneath the various troubles of pesticide residues, industrial wastes, vanishing wildlife, and dwindling wilderness. It gave people a way to think about the roots of environmental problems, provided a measure for our action and a goal to aim for. It spoke to individual love of nature, but moved from nature as a part of life to our obligations in life toward nature. The land ethic was America's greatest intellectual contribution to West-ern thought about humans and nature, and environmentalists have only begun to plumb its depths.

By the 1940s a few people had built a genuinely spiritual vision of humans and the land around the perspectives of ecology and evolution, and Aldo Leopold made a serious moral framework for our treatment of nature, but it took time for that view to spread through the culture. Leopold's outlook took shape in the late 1930s, and during World War II he used the leisure enforced by gasoline rationing and a lack of graduate students to write key essays like "Thinking Like a Mountain" (1944) and to plan what became *A Sand County Almanac*. By then the National Park Service was protecting predators and eliminating nonnative species from the parks, and nature writers had moved beyond focusing on organisms and individual actions to frame narratives around the processes and rela-tionships that sustained life on the land. In 1940 Donald Culross Peattie wrote what Leopold hailed as the first ecological novel, *A Prairie Grove*, and a year later Rachel Carson, then a U.S. Fish and Wildlife Service biol-ogist, published a book on the ocean's life, *Under the Sea Wind*, that aban-doned the human viewpoint for that of the ocean and its creatures.

A Sand County Almanac appeared in 1949, a year after Leopold's death, to widespread and enthusiastic neglect, but as awareness of environmental problems increased, it became popular. Publishers brought out cheap paperback editions—including Ballantine's, still in print, in the Style of Life series—and people started quoting Leopold with a reverence appro-priate to scripture. No one erected altars to evolution or ecology or saw them as conscious agents of our fate, but the knowledge that we live in a world shaped by forces that go back to the beginning of life and hold it in being each moment gave a new significance to what we saw. The bird's wing and the spider's web revealed ancient processes; the sprouting brush and small trees in an abandoned field pointed to plant succession and changes in animal life; and a patch of snow swept by a wing and stained by a bit of bloody fur recorded the forces that tied species together. The mechanisms of evolution and the processes of ecology became the visible hands of the nonexistent God.

Consciousness of connections grew as people came to understand how much humans threatened the existence of the independent, endless natural world earlier generations took for granted. "The end of the unknown is at hand," Leopold sadly wrote, and went on to wonder if that experience could be lost without "something likewise being lost from human character."[70] He was glad that he would "never be young without wild country to be young in. Of what avail are forty freedoms without a blank spot on the map?"[71] In the 1920s he had begun thinking about preserving wild country, recommending to the Forest Service that some parts of the national forests be set aside for wilderness values, and in 1935 helped found the Wilderness Society, which, after World War II, campaigned for a national wilderness system. The wilderness movement provided a bridge between conservation and the environmental movement. It became the emotional center for nature protection, a political campaign for sacred space, the first and in many ways the most serious expression of the religious impulse behind the calls for pollution regulations and new land use regulations that formed the new environmental movement's visible public agenda in the late 1960s. The rise of wilderness is the next part of our story.

3 / Journey into Sacred Space

"Wilderness," like "nature," has many, contradictory meanings, and Americans have used them all and added a few of their own. The European settlers arrived with biblical ideas of wilderness—as exile and the Devil's haunts, the place where God met and led the people of Israel, the prophets took refuge from their enemies, and Christ prepared for his ministry—and with European ones—of wild lands as the haunt of wild beasts and wild men—and they looked in the wilds of America for fabulous kingdoms and unknown animals. After independence they came to see the land as an endless storehouse where all could prosper. By the end of the nineteenth century they believed the encounter with wilderness had shaped the national character. They also found in wilderness, following Muir, the doorway to reality and a refuge from society. The generations that followed fixed on landscapes without visible evidence of industrial society as the place for a spiritual journey. Here, where the veil between human society and the world's realities thinned, those who abandoned modern technology and immersed themselves in the land could come to an emotional understanding of the intellectual truths of our ties to nature. Wilderness became a key conservation cause by the 1960s, and a decade later a geographer described it as "a contemporary form of sacred space, valued as a symbol of geopiety and as a focus for religious feeling."[1]

If it seems too much to call wilderness sacred or to see wilderness journeys as spiritual quests, go back to the reactions people have to everyday nature. Each fall, the leaves of New England's trees turn from green to

yellow and red, first individual leaves, then a few trees, then entire mountainsides. That predictable and repeated spectacle brings out enough
people, most of them otherwise insensitive to "nature," to make "leaf season" a distinct peak in the New England tourist industry, with buses and
festivals and "leaf-peepers'" dinners ("6 P.M. Saturday night at the town
hall," as the sign on one village green in central New Hampshire advertised). Thousands seek out canyons in the Southwest where only the wind
and an occasional bird cry break the silence, and the only sights are sand-
stone and sky. Public interest made photos of the canyons into a minor
industry. Even without the spectacle, nature affects people who have never
read Emerson's essays. The kids in my basic-training platoon took the ser-
geant and the casual violence of the training cycle in stride, but on a field
exercise, off the road, looked with awe at the Southern pines around us.

People who feel pulled to the wild find enchantment in even small
breaks from the social world. In a northern winter snowstorm, a few steps
into woods on snowshoes or skis takes you to a place where there is only
leaden sky, spruce trees, falling snow, and fallen snow, and the only sounds
the hiss of falling snow or blowing crystals. If you are one of those who
could not live without wild things, what price in extending that isolation,
being where miles stretch away without any human structures, where no
one has ever systematically cleared the land of its plants or killed most of
the animals or bulldozed the ground, and where wild lives and wild communities go on without humans? What value in just knowing that world
still exists? Unless we define the sacred as something beyond this world,
wilderness functions as sacred nature for millions, the site of experience
or an anchor for hope.

While Americans embraced nature in various forms from the first days
of settlement and gave wilderness many meanings, they never gave it a
permanent place. It was raw material for civilization, land that should be
turned into farms and pastures as quickly as possible, and even when they
set areas of land aside for their natural qualities, they did not include
wilderness. The national parks preserved monumental scenery and geological wonders that visitors could view in comfort—not primeval landscapes. Only in the 1920s, as roads and tourist lodges approached the last
large roadless areas in the continental United States, did a few people ask

if wild land could have or should have a place in industrial civilization. Over the next two decades a few worked within the Forest Service to make wilderness a land use, and in 1935 a group established the Wilderness Society to carry on that campaign and to foster discussion of the value of wilderness.

After World War II, as increased leisure brought a boom in outdoor recreation and increased development threatened the lands people wanted to see, wilderness became a popular cause. By the time Congress passed the first wilderness protection act, in 1964, wilderness had a prominent place in conservation's political program. By then it also had new meanings. The men who founded the Wilderness Society drew on Muir's search for spiritual goods in the wilderness and on the nostalgia for the pioneers and their virtues that accompanied the rise of industrial, urban America. After the war their successors added science as an argument for preservation. From ecology's perspective, wilderness could be seen as land where natural ecosystems remained largely intact and functioned without significant human interference, land that not only looked wild, but biologically lay beyond society.

Through these changes Emerson's legacy remained. People accepted his belief that "the spiritual is not a realm apart from the natural but is instead revealed—and alone revealed—through the natural," and it made no difference that they now defined the natural in terms of the presence of top predators, for that meant the ecosystem was intact.[2] Wilderness became a popular cause because it was the place where people might search for insight into the world. That human development had threatened it only made it more precious and its defense the more important.

Tracing the rise of wilderness and its links to a spiritual search from the early twentieth century to environmentalism involves following two linked movements: one away from Romantic ideas of nature toward an "objective" and "scientific" view, and another away from a focus on individuals developing pioneer virtues in a struggle with the wild to people seeking insight in what they regard as a sacred space. Between 1920 and 1940, as wilderness advocates were discussing wilderness as a land use and, in the Forest Service, building a system of administrative protection for large areas, ecologists were coming to an understanding of ecosystems that allowed people to see wilderness as nature in its purest form, the world as it "should be" (scientists, it should be said, built into their theories the ambiguities and value judgments involved in saying what ecological relationships "should" be and identifying these as a measure of good). In the 1950s, as the Wilderness Society's campaign for statutory protection for wilderness grew, Sigurd Olson made the wilderness journey a spiritual pilgrimage, but even as books like *The Singing Wilderness* (1956) drew people to union with nature, a darker view developed, a perspective closer to Robinson Jeffers than Emerson.

With the Wilderness Act of 1964 a new period began, with wilderness's defenders taking political and administrative action, and some, frustrated at the law's slow pace, using sabotage to defend threatened areas. Debate about the definition of "wilderness" and the relevance of the concept to our lives continued, the argument at times developing in ways some felt betrayed the cause. William Cronon's suggestion, in "The Trouble with Wilderness," that the wilderness concept did not always serve the long-term interests of the wilderness movement brought a hail of criticism, but Cronon was not alone. A recent collection of papers had the apt title *The Great New Wilderness Debate*.[3] The debate was intense, not only because of people's passionate commitments and the threat to wilderness, but because wilderness, linking the material and the transcendent, held science and spirituality in tension. The stakes were all the greater because wilderness seemed a place to explore and possibly heal divisions deep in modern society—the alienation of body from mind, observer from object, and humans from the world—and bring about a union of thought and action, emotion and reason.

A PLACE FOR WILDERNESS AND WILDERNESS PLACES

From the early nineteenth century, Americans looked on wilderness as something destined, like the forests, wildlife, and the native peoples, to pass away before the triumphant progress of civilization, and even when they set aside areas of nature as parks, they ignored wilderness in favor of Romantic scenery. The early national parks had majestic mountains, enormous canyons, and beautiful waterfalls (and in Yellowstone's case geysers and mudpots as well), and the National Park Service (NPS) worked to make it easy for Americans to see the parks' wonders in comfort. It was not, in fact, the NPS, responsible for the parks, that first experimented with wilderness as a land use, but the United States Forest Service (USFS), a distinctly utilitarian agency that managed public land for multiple uses and measured success by dollars from timber sales. After World War I, noticing the public's growing interest in outdoor recreation and the rise of the National Park Service, it asked each district to assess its lands for public use. Aldo Leopold and Arthur Carhart, surveying the southwestern district, made the startling suggestion that some areas be left as they were for their wilderness values.

For this radical change, they offered a conventional argument: the appeal of recreation that re-created pioneer life. Leopold viewed wilderness as "a continuous stretch of country preserved in a natural state, open to lawful hunting and fishing, big enough to absorb a two weeks' pack trip, and kept devoid of roads, artificial trails, cottages, and other works of man."[4] Setting aside these areas "would not subtract even a fraction of one per cent from our economic wealth, but would preserve a fraction of what has, since first the flight of years began, been wealth to the human spirit."[5] Carhart, transferred to the upper Midwest, made the same plea for what became the Boundary Waters Canoe Area along the Minnesota-Ontario border, though the great attraction was not pioneer hunting but the romance of the voyageurs. Travelers could paddle their canoes along the old fur trade routes, paddling like the French-Canadian and Hudson Bay men, fishing and camping. The Forest Service set aside the first wilderness, in the Southwest, in 1924, but limited protection to an administrative declaration that could easily be revoked if someone wanted to cut timber or dig a mine.

In 1925 Leopold warned it would be a mistake to think that because a "few foresters have tentatively formulated a wilderness policy, that the preservation of a system of wilderness remnants is assured. . . . The Forest Service will naturally select . . . the roughest area and those poorest from the economic standpoint. But it will be physically impossible to find any area which does not embrace some economic values. . . . Unless the wilderness idea represents the mandate of an organized, fighting and voting body of far-seeing Americans," wilderness would disappear whenever a private interest really wanted to use an area.[6] All wilderness, he recognized, could disappear. Civilization had been an island in a sea of wilderness—even in nineteenth-century America wilderness had loomed large. Thoreau had looked toward what seemed a largely unexplored West, and Muir had found the Sierra apparently pristine. Now people so thoroughly surrounded wild nature that wild areas would survive only if people consciously chose to preserve them, and that meant wilderness had to have a place in their lives. They might defend a relic of the past or an aesthetic delight for a time, but sentiment would dwindle with the passing years.

To build such a group, eight men formed the Wilderness Society in 1935. Working for a dream, they were more than dreamers. Leopold, Carhart, and Robert Marshall (in the early years the group's most active member and its de facto leader) had administered land for government agencies, and the others had comparable experience. They talked about pioneer recreation and character-building and focused on humans in the wilderness and wilderness for our use, but wilderness's deepest appeal was as another world. The photographs of Ansel Adams, a Sierra Club member and activist, showed that. Adams began working in the interwar years, and his black-and-white photographs made him as influential as the Hudson River School had been a century before. People hung prints on the wall and displayed the Sierra Club's coffee-table books in much the same spirit and with the same fervor as immigrant Catholics placing the Sacred Heart of Jesus on the dining room wall. Adams's images appeared on postcards, place mats, and note cards—and recently as screen savers. The photographs that made Adams famous showed rock, trees, and wild country—only rarely people's works, even more rarely human figures. People occasionally criticized his pictures "as being inhuman. 'Doesn't anybody go there?' asked art critic Lincoln Kirstein" after seeing an Adams exhibit.[7] An entire school of landscape photography grew up around his work, and even when his disciples, such as Eliot Porter, used the lusher medium of color film, they still showed a world apart—spiritual solace, not home.

Adams's photos gave form to a view deeply embedded in American culture, so deeply that even scientific ideas showed its impress. Ecologists, as committed as any group of scientists to the idea of objectivity, based their early research in this romantic vision of nature as a world apart. Frederic Clements's theory of plant succession, for instance, which dominated plant ecology and then range management during the first half of the twentieth century, separated people from nature as thoroughly as Emerson divided Soul from Nature. Clements described a harmonious world in which organized plant communities succeeded each other in strict order.[8] Reflecting the botanists' love for their subjects, the theory classified things that destroyed plants—people, grazing animals, even fire—as outside nature's system, "disturbances" to the plant community.

The reaction of British ecologists to American definitions showed this was a cultural quirk, not a matter of data and observation. In 1923, the English ecologist A. G. Tansley remarked of Clementsian theory, with its fondness for plant communities, that the "belief is sometimes met with that only perfectly 'natural' vegetation is a proper subject for ecological study. If this were true, the ecological field in Great Britain would be very limited indeed."[9] British plant ecologists never adopted Clementsian ideas or perspective, and their counterparts in animal ecology included humans and their landscapes as a matter of course. Charles Elton, whose *Animal Ecology* (1927) guided research in that field for a generation and whose research group attracted ecologists from around the world, did most of his research in the fields, pastures, and hedgerows of Britain, and he insisted his students work on problems of economic importance. The British ecologists' rejection of Clements's theory suggests the American view of people and nature came more from national experience than any "objective" or "scientific" stand. That the American scientists, though committed to "objective" and "scientific" views, smuggled this idea, all unwittingly, into their theory, suggests how deep the Romantic current ran in American society. The persistence of this separation between humans and nature in the concept of sacred wilderness and the belief in the wilderness journey as pilgrimage—views that helped bridge the movement from conservation to environmentalism—shows its continuing power in the culture.

MAKING WILDERNESS SACRED

In the middle of the twentieth century Sigurd Olson carried the wilderness cause, seeking a connection to the universe in the wilderness, and in his books establishing wilderness for many as sacred space and the wilderness journey as a pilgrimage. Though Olson spoke of pioneer virtues and the romance of canoeing, he emphasized not activity but openness to experience, and in his hymns to the canoe country along the Ontario-Minnesota border told of a wilderness beyond Romantic scenery: "The intangible values of wilderness are what really matter, the opportunity of knowing again what simplicity really means, the importance of the natural and the sense of oneness with the earth that inevitably comes within it. These are spiritual values. They, in the last analysis, are the reasons for its preservation. Wilderness will play its greatest role, offering this age a familiar base for exploration of the soul and the universe itself."[10]

He loved a hard land, the southern part of the Canadian Shield, that mass of Precambrian granite running from Hudson Bay into the upper Midwest. Not covered with soil, it could not be farmed and so was not settled, and it formed such a formidable barrier to travel that until the railroad came canoes were the only good way to cross it. The Group of Seven, the first successful and influential Canadian landscape painters, made it the iconic Canadian landscape in the early twentieth century. They presented a country of lakes, pine and spruce forests, thin soil, and hard winters, the very image of the unconquerable North. Olson loved it from his college years, lived to travel in it, as an adult worked to save it as wild country, and in the late 1930s began writing about it. In the 1950s he received honors from conservationists and acclaim from the public.

Born in 1899, the son of a Swedish Baptist minister, Olson grew up in parsonages in Wisconsin, and, like Emerson, Muir, Burroughs, Seton, and others, found in nature solace from the chilly Protestantism of home. At the age of seven, alone on a pier in Lake Michigan, he heard for the first time what he called the song of the wilderness: "That day I entered into a life of indescribable beauty and delight."[11] He heard the singing in many places, but "the possibility of hearing the singing wilderness and catching perhaps its real meaning" seemed greatest "in the wilderness lake country of the Quetico-Superior, where travel is still by pack and canoe over the ancient trails of the Indians and voyageurs."[12] He called others to the quest, for our attempt to satisfy "emotional needs that were ours before the dawn of history" was common to humanity.[13]

In the 1920s, working as a guide on canoe trips in the border country of Minnesota, Olson found "that sense of oneness which comes only when there are no distracting sights and sounds, when we listen with inward ears and see with inward eyes, when we feel and are aware with our entire beings rather than our senses. . . . Without stillness there can be no knowing, without divorcement from outside influences man cannot know what spirit means."[14] His biographer, David Backes, spoke of Olson's "obsession with what he called his search for meaning," but "obsession" is too strong a term.[15] Olson searched not because he was obsessed, but because he was a mystic, an ordinary, if somewhat uncommon, human type.

Olson's search for meaning in life led him to write, and while his first piece, "Why Wilderness?" (1938), emphasized healthy recreation, friendship, and a better perspective on life, by the 1950s he described wilderness as the gateway to the transcendent and the wild as the source of deeper meaning. He pushed beyond individual spiritual experience to defend wilderness as a crucial spiritual component of civilization. By then, he also set out to organize his "theology," a term Backes felt was appropriate because Olson's philosophy "was deeply connected with his beliefs about the nature of God."[16] He could not, as Muir had, finesse the separation between spirituality and science by selectively appealing to scientific disciplines and using the language of natural theology.

Science had too strong a place in the culture and was too unified for the first strategy to work, and natural theology had no appeal, which precluded the second. Instead, Olson used science to describe the land, even suggesting that our evolutionary heritage explained our attraction to nature, but he rejected the materialism often associated with science—the belief that there was nothing in the universe but matter. That materialist perspective, he thought, threatened to destroy "the religious truths that had provided spiritual sustenance, the truths that had given meaning, and offered nothing in their place."[17] He believed there were other truths beyond science, of equal or greater value, treasures of the spirit, accessible to the individual, found in the wilderness, and part of a fully human life.[18]

To frame those truths he drew on a line of modern thought that explored spirituality outside conventional creeds. He read Lecompte du Nouy, whose *Human Destiny* presented evolution on the "spiritual and moral plane" (de Nouy's phrase), and Teilhard de Chardin, who proposed that evolution formed a new level of the world (the sphere of consciousness) where all evolved toward union with God (his Jesuit superiors and the Roman Curia found this alarming, and Teilhard spent several decades doing paleontological work in Manchuria). From Lewis Mumford's *Conduct of Life* Olson drew the idea of an evolving Godhead.

His views were probably closest to those of Aldous Huxley, who argued, in *The Perennial Philosophy*, for a common ground in all religions beneath the level of creeds, a "metaphysic that recognizes a divine Reality substantial to the world of things and lives and minds; the psychology that finds in the soul something similar to, or even identical with, divine Reality; the ethic that places man's final end in the knowledge of the immanent and transcendent Ground of all being."[19] For Huxley, individual experience served as "the focal point where mind and matter, action and thought have their meeting place in human psychology." He believed that the "divine Ground of all existence is a spiritual Absolute, ineffable in terms of discursive thought but (in certain circumstances) susceptible of being directly experienced and realized by the human being." Experience was primary; theological formulae failed in the face of the thing experienced. Both were searching for insight, but Olson rejected Huxley's search for enlightenment through meditation, looking instead to wilderness.[20]

Into the mid-1960s *The Singing Wilderness* sold twice as many copies as Leopold's *Sand County Almanac.*[21] The book's appeal was neither literary nor intellectual. Olson's prose was serviceable at best, and his ideas were more heartfelt than original; however, he wrote of silence and solitude found paddling on a wilderness lake or sitting around a campfire, moments of peace and oneness with the universe. For a generation that remembered but had missed the last days of woods travel and work without power tools he framed travel in terms of pioneer nostalgia; to people who believed in science but also in spiritual values he wrote of the land in terms of science and sentiment. He pointed to a high but intelligible goal: saving wilderness as "a stepping stone to cosmic understanding. In a world confused and strident, where all the old verities are being questioned, this is the final answer." As Backes pointed out, "When the *primary purpose* of wilderness is defined in essentially sacramental terms, then the wilderness itself becomes a sacred temple, and anything that either violates the temple or disrupts the 'service'—the cosmic communism at the heart of a wilderness experience—is out of place."[22] Olson made conservation a moral duty.[23]

The postwar movement stood between nineteenth-century nature appreciation and a conception of wilderness that developed in the 1990s. Olson defended what we would recognize as wilderness and the spiritual values of contact with the truly wild, but described wilderness journeys in terms of what the pioneers did. He used evolution's perspective, but (necessarily) lacked the environmental understanding of nature as complex webs of relationships. The wilderness movement existed in a similar state, using ecological terms to argue for wilderness, but relying on boundaries on the land to protect wilderness from development.

In the late 1960s, circumstances changed. The growing understanding of ecological relationships and of the pervasive impact of humans and human technology forced the wilderness movement to grapple in new ways with the relation between wilderness and society. Wilderness areas, it seemed, were deeply intertwined with the biotic communities that directly supported human society every day and in every place. Wilderness was less a place apart, for individual contact with the ultimate, than part of our own world. The kind of wilderness people valued also changed, and a new debate began, one still changing our ideas and appreciation of wilderness.

DANGEROUS WILDERNESS

The Wilderness Society began with fewer than a dozen members, and when Howard Zahniser began his campaign for congressional protection for wilderness in the late 1940s only a few joined. Over the next two decades wilderness attracted many followers, became an environmental issue, and acquired new meanings and values without fully shedding the old. Change began with the postwar economic boom which gave people more time to seek out "unspoiled" nature while destroying it at a faster rate. What a few in the 1930s saw as a looming possibility—the end of large areas of wild country—became by the 1950s an imminent and visible threat. Nature protection organizations changed their missions to meet the new situations. The Sierra Club, for example, removed from its bylaws the phrase about improving access to the mountains and worked, instead, to control development. It, and other groups, reacted most strongly to the Bureau of Reclamation's plans, announced in 1950, for a series of dams on the upper Colorado River. One of them, at Echo Park, would form a reservoir that would flood Dinosaur National Monument. That roused memories of Hetch-Hetchy Valley in Yosemite National Park, lost in 1914 to a dam for San Francisco's water supply.

The battle over Echo Park changed the politics of nature protection, making the Sierra Club and its charismatic executive director, David Brower, important national figures and revealing changes in ideas about what we might find in wilderness. Wilderness advocates had campaigned for areas that allowed only primitive travel but not threatened the travelers. The Colorado canyons and the plateaus around them presented more pain than the discomforts of pioneer life and less beauty than the usual Romantic scenery. The land, though spectacular, lay far enough beyond the Romantic aesthetic to be an acquired taste; plants were few, small, and low, and animals, scattered and hidden.[24] It showed the bones of the world, masses of sedimentary rock, sculpted by wind and rain. The land had little rain and no streams. Discomfort and even danger were constant. During the day, temperatures rose to dangerous highs, then fell quickly at night. A hot day and too little water, even on an easy hike, or a broken-down car off the main roads meant serious trouble. Hiking in the backcountry required planning, care, and experience. That the canyons, which had drawn a trickle of admirers, now had many defenders, suggested a new view of nature and wilderness had become popular.

The loss of Glen Canyon, unknown and overlooked to the end, fixed the canyons in the imagination of wilderness's defenders. The bill that spared Dinosaur National Monument provided for a dam at Glen Canyon, and before the waters rose David Brower and some other of Dinosaur's defenders floated down that section of the Colorado. They found it incredibly beautiful, and the doomed Glen Canyon became a focus for frustration and one of conservation's lost causes.[25] Edward Abbey later wrote a novel, *The Monkey Wrench Gang,* about a plot to blow up the dam. As the waters filled behind Glen Canyon dam, the Sierra Club issued a book entitled *The Place No One Knew,* seventy-five of Eliot Porter's photographs of the area with texts speaking of the spiritual values of wilderness.

Mourning lost beauty, it rallied people even more effectively than Wallace Stegner's *This Is Dinosaur,* a Sierra Club book with pictures and text that described Dinosaur National Monument at Echo Park.[26] Stephanie Mills, a bioregional activist, found in *The Place No One Knew* a vision "of pristine and unimaginable beauty" and a "clear invitation to make common cause with the planet, its wilderness especially." She had many good reasons to identify with the planet, but "what I think I was yearning after was the uninventedness of that life and beauty—its perfection absent thinking, strife or agency; its suchness and freedom from the need to improve." The book's "philosophy of relationship to nature . . . made perfect sense to me: nature, epitomized and embodied in wilderness, was unsurpassable by any human artifact."[27]

The Place No One Knew may have presented a new "philosophy of nature," but it did so using familiar terms. Though it looked on nature as sacred and its defense as a holy cause, it viewed wilderness as scenery—indeed, the kind of scenery that filled the national parks. Porter's photographs stood in a well-established aesthetic line, the visual presentation of Romantic sentiments in monumental scenery. They showed soaring masses of stone, quiet pools, plants against stone cliffs, delicately picked out areas of color or great swaths of changing shades. The texts called on us to preserve wilderness for its spiritual values and to seek the transcendent in nature—ideas that went back to Emerson. Even as an environmental statement, *The Place No One Knew* took no radical stands. Brower called for "reasonable development" and suggested that the coal of the Upper Basin states would be a more reliable energy source than hydropower. A decade later these sentiments seemed out of date. This coffee-table art book, though, introduced into the wilderness debate something new, a sense of sin and collective responsibility. "Glen Canyon died in 1963," Brower wrote, "and I was partly responsible for its needless death. So were you. Neither you nor I, nor anyone else, knew it well enough to insist that at all cost it should endure." And this ignorance "has cost all men, for all time, the miracle of an unspoiled Glen Canyon."[28]

The book guided readers from beauty to insight. The preface to the first section, "The Place," described a continuing creation. One river built the stone of the canyon walls; a second, the modern Colorado, shaped it. "You didn't quite catch the river in the act of sculpturing, but the color of the Colorado assured you that creation was still going on."[29] The text harked back to Muir: "When your spirit cries for peace, come to a world of canyons deep in an old land."[30] The second section, "The Idea," declares: "We shall seek a renewed stirring of love for the earth; we shall urge that what man is capable of doing to the earth is not always what he ought to do; and we shall plead that all Americans, here, now, determine that a wide, spacious, untrammeled freedom shall remain in the midst of the American earth as living testimony that this generation, our own, had love for the next."[31] Love required more than land use programs or an appreciation of American history. It called for spiritual development.

A quotation from William O. Douglas said that "to be in tune with the universe is the whole secret," and one from Aldo Leopold ended with his declaration that the task of recreational development is not building roads into lovely country, but building "receptivity into the still unlovely human mind." A passage from Albert Einstein described the mystical "the most beautiful and most profound emotion we can experience. . . . To know that what is impenetrable to us really exists, manifesting itself as the highest wisdom and the most radiant beauty which our dull faculties can comprehend only in their most primitive forms—this knowledge, this feeling is at the center of true religiousness."[32]

Meantime, wilderness's believers moved toward a new view of nature and its relation to humans, more complex and less comfortable than the old, emphasizing pain, danger, difficulty, and failure more than courage or insight. In the older literature of wilderness travel endurance brought reward.[33] In the new one it might, but it might not. Edward Abbey became the American desert prophet by preaching this harsher gospel. He sneered at scenery. Looking did nothing, he claimed. Leave your car and walk, he said, or better yet, crawl. Leave all, sacrifice all, strip the self bare, and come to the last bit of energy and hope: "When traces of blood begin to mark your trail you'll see something, maybe. Probably not."[34] Muir had found fellowship in the mountain, fancied "a heart like our own must be beating in every crystal and cell, and we feel like stopping to speak to the plants and animals as friendly fellow mountaineers."[35]

Abbey celebrated the desert world precisely because it was ahuman. He went there not only to escape mechanical civilization, but "to confront, immediately and directly if it's possible, the bare bones of existence, the elemental and fundamental, the bedrock which sustains us. . . . I dream of a hard and brutal mysticism in which the naked self merges with a nonhuman world and yet somehow survives still intact, individual, separate." So far, Thoreau in a really bad mood, but Abbey went on: "The desert says nothing. Completely passive, acted upon but never acting, the desert lies there like the bare skeleton of Being, spare, sparse, austere, utterly worthless, inviting not love but contemplation. In its simplicity and order it suggests the classical, except that the desert is a realm beyond the human and in the classicist view only the human is regarded as significant or even recognized as real."[36]

The Sierra Club's Publications Committee dealt with changing views of wilderness when it considered the proposal to publish *Not Man Apart,* a book of photos of the Carmel–Big Sur coast accompanied by pieces of Robinson Jeffers's poetry. On the surface the idea seemed unexceptionable.[37] Jeffers's poetry inspired many, including Ansel Adams, and the area was undeniably beautiful. But, as Michael Cohen pointed out in his history of the Sierra Club, "the introduction of Jeffers into the Club canon made it difficult to avoid a more complex and comprehensive critique of American civilization." Jeffers was not a classical humanist, and "the Club feared to make its argument for wilderness in terms which went beyond humanism. The humanistic argument was the heritage of Robert Marshall and Howard Zahniser. At the same time, Loren Eiseley, who wrote the introduction to *Not Man Apart* and whom Brower hoped would become a new and important voice for the Club, represented a newer, colder view of nature, one increasingly present at the wilderness conferences, in which wilderness was not pretty but a world of mindless evolution. Eiseley's view accepted, as the poetry of Jeffers did, the violence, death, chaos, and change of environmental history, where man himself was frighteningly alone and acted often as an animal. . . . At the same time, then, that the Exhibit Format books were representing nature idealized and picturesque, the avant-garde of the Club was moving beyond that kind of thinking."[38]

"Classical humanism" emphasized intellect and the contemplation, appreciation, and celebration of nature. Jeffers looked steadily at the irrational elements in humans, particularly their primal passions. His long narrative poems, often set on the California coast, used the themes and actions of Greek myths and revolved around murder, incest, and the passions that produced them. Instead of the nobility of human thought, he wrote of "the bitter weed / of consciousness." The "learned astronomer / Analyzing the light of most remote star swirls" found the stars fleeing. No doubt that was to escape "the contagion / Of consciousness that infects this corner of space" ("Margrave"). Jeffers seemed to find the very numbers of humans a blight, imagining the storm that would "scour more than coast lines" and "the cities gone down, the people fewer and the hawks more numerous, . . . the two-footed / Mammal, being someways one of the noble animals, regains / The dignity of room, the value of rareness" ("November Surf"). He found the good in humans not in their reasoned appreciation of the world, but in their working within nature's systems and with its rhythms. Even here Jeffers ran against conservation sentiment, which accepted humans within the rhythms of nature but said nothing about earning one's living, in nature or out of it.

The Club assimilated parts of this harsher view of nature, but only parts. It dealt with Jeffers by selection, taking appropriate sentiments from his work and ignoring some of the implications. For the photo book on the Carmel coast the Club used Jeffer's phrase "not man apart" from lines declaring that "the greatest beauty is / Organic wholeness, the wholeness of life and things, the divine beauty of the universe. Love that, not man / Apart from that," choosing to emphasize the wholeness of the world and human ties to it rather than Jeffers's concern with the magnificence of the world and the insignificance of humanity. Even bowdlerized, Jeffers

remained a minor enthusiasm, for only the completely ignorant or perfectly enlightened could be fully comfortable with his unblinking view of
life's pain and the insignificance of the human race in an indifferent universe. Loren Eiseley, who "accepted and celebrated, as Jeffers's poetry did,
the violence, death, chaos, and change of environmental history, where
man himself was frighteningly alone and acted often as an animal," gained
a large following, but he used accepted literary devices and sentiments to
blur the edges of despair. He set his ideas within familiar narratives: nostalgia for youth and the hardships of Depression-era America, the young
man discovering the world of the intellect, and, most important, the opposition between human sentiment and sensibility on the one hand and the
hard objectivity of scientific research on the other.

DEFENDING WILDERNESS

Events as well as ideas shaped discussion. The Wilderness Act of 1964 established wilderness as a federally defined and protected land use, set aside
some areas, and put in place a legal mechanism to add others to the system, but many considered its provisions too little, too late. In the next
few years, rapid development and little protection made the situation more
alarming. Roads split large areas into ones too small for wilderness protection, oil and gas drilling and mine claims ruined others, and reviews
of roadless areas brought only more reviews and comments, followed by
protests and lawsuits. In 1981 a group of frustrated wilderness lovers—
some disillusioned lobbyists, others alarmed hikers—formed a new
group, Earth First!, dedicated to an uncompromising defense of wild lands.

Earth First! came to the public with a piece of guerilla theater that
entered environmental folklore (on Earth Day 1983 a team unfurled a 300-
foot-long piece of canvas with a crack painted on it down the face of Glen
Canyon Dam) but stayed a center of controversy for its association with
sabotage to protect the environment. Direct action to protect wilderness,
which became known as "ecotage," took as its model and patron saint
Edward Abbey, who had, in *Desert Solitaire,* told of tearing up the survey
stakes for a new backcountry road in the park he worked in, and had
opened *The Monkey Wrench Gang* with a description of burning down

billboards in the desert detailed enough for the interested reader to use as instructions.

For a decade Earth First! occupied headlines and attention inside and outside the environmental movement.[39] Taking seriously its motto, "No compromise in defense of Mother Earth," the group skated close to the law's edge by "recommending" or just "describing" such actions as spiking trees, crippling earth-moving equipment, and destroying logging roads, justifying such action by pointing to the sacredness of nature and the immediate threat to our last wild places. Although the group called for measures against property not people and warned that successful ecotage would not kill or injure people, critics accused it of indifference to human life. Opposition reached a high point in 1987, when pieces of a shattered saw blade injured a mill operator in California. Earth First!'s enemies demanded prosecution and imprisonment. Ecotage also attracted the attention of federal authorities, and an FBI sting operation resulted in the arrest of several people for the attempted dynamiting of a power pylon. It could hardly have been otherwise; a society that regarded property as the defining characteristic of the individual and economic development as the central social good could only see Earth First!'s program as an attack on core values.

Success as well as opposition and prosecution shaped Earth First! Publicity brought new members, who diluted the founders' influence, and added new and—some thought—irrelevant issues. Radical stands divided members against each other and, at times, discouraged supporters. One article in the Earth First! magazine analyzed AIDS as a way to reduce the population; another recommended ending all immigration into the United States. Dave Foreman, one of the founders, left, charging that Earth First! took too many stands not related to wilderness and ones that undermined its appeal to ordinary Americans.[40] He went on to edit *Wild Earth*, and Earth First! survived, if in somewhat different form. So did radical action. Despite prosecutions, people continued to spike trees and drain the crankcases of bulldozers in the back country. No one could say, obviously, just how many people joined in or how many fires blamed on environmental terrorism had been set for revenge or insurance, but interest was high enough to produce a second edition of *Ecodefense: A Field Guide to Monkeywrenching,* one with enough changes and new tips to suggest people had used the first.[41] A crude product, it was still deadly serious and grounded in principles—a shop manual for sabotage with philosophical justification included.

Behind the fervent physical defense of wilderness lay an equally strong belief in wilderness as an ultimate value. People give many reasons for defending these lands; a recent summary of "the traditional and contemporary arguments for wilderness" lists thirty, including arguments based on physical recreation, mental inspiration, intrinsic value, conservation, preservation, nature appreciation, economics, ecology, and the counterculture.[42] But people don't risk fines or jail or their lives because something might evolve in the next few million years or because wild land forms our nation's character. Julia Butterfly Hill, for instance, sat 180 feet up in a redwood tree for 738 days, and neither common sense nor her testimony suggested economic or utilitarian motives or even a particularly well-developed intellectual defense of nature or wilderness.

She had acted, she said, because clear-cuts were turning "these majestic ancient places which are the holiest of temples, housing more spirituality than any church," into muddy fields.[43] In *Confessions of an Eco-Warrior,* Dave Foreman declared that wilderness has value because it "*is.* Because it is the real world, the flow of life, the process of evolution, the repository of that three and a half billion years of shared travel." All things are good and equally good "because they exist." The preservation of wilderness is "an ethical and moral matter. A religious mandate. Human beings have stepped beyond the bounds; we are destroying the very process of life."[44] That conviction provided the passion beneath the wilderness movement.

RETHINKING WILDERNESS

The reaction to William Cronon's essay "The Trouble with Wilderness" showed how deep people's belief in the reality of wilderness ran. Cronon did not attack wilderness; he made an academic argument, growing out of a semester-long seminar on nature and culture, about the ways we see the world and how our ideas shape our actions. While reality is independent of human perceptions and belief, he said, we see the world only through these filters. We have, that is, no God's-eye view; we approach the world with expectations formed by our culture and education. For that reason we should realize that "wilderness" is a word we give to a part of what is "out there." He emphasized that this did not make wilderness less real, less interesting, or less worthy of our respect, even devotion. Certainly it did not in his case, for he pledged his allegiance to environmentalism and believed wilderness was important and that large wilderness areas should be preserved.

He did warn that because wilderness was a concept rather than a reality, an exclusive focus on these "undisturbed" areas tempted us to see wilderness as something beyond society rather than a continuing presence in our lives as part of our daily surroundings. The trouble with wilderness, then, was that the concept encouraged us to neglect our lives as we live them and as we ought to live them in favor of an idealized realm "out there" that we visited or dreamed about. We should not discard the idea of wilderness, Cronon said, but we should fit it into an inclusive view of how we live with and for nature.

Cronon's stand raised no eyebrows in Western philosophy, which readily acknowledges that the ideas in our heads do not exactly match the world outside them and that what we know affects what we see. That does not mean that what we construct in our heads has no reference to the world outside or that there is no world outside. Philosophers agree that we see as through a glass darkly, but there is something out there to be seen and we see it. That nature includes more than wilderness also seems unexceptionable. Besides the philosophers, the writers and naturalists who formed American ideas about nature believed—indeed insisted on—a broad definition of nature and, before the twentieth century, did not think of wilderness as a thing apart.

Cronon's essay, though, "started a firestorm of debate."[45] The audience at the American Society for Environmental History convention suggested that Cronon gave an unbalanced critique or played into the hands of the anti-wilderness coalition. Two of the three comments in *Environmental History* (mine was the third) argued against the essay. Michael Cohen thought Cronon's analysis led to thinking of wilderness as something "textual" rather than real. "Wilderness," he said, "can be reduced to a social construction using the same methods of analysis that reduce mountains to cathedrals. But mountains are not made by men." Cohen ended with a ringing declaration of the importance of "first nature" (the actual world beyond human society) in the American West and an inclusive view of our ideas about wilderness. Samuel P. Hays, drawing on his own experience in the eastern wilderness movement, argued that Cronon paid too much attention to ideas about nature, too little to people's motives. Wilderness advocates, he said, do not think of creating a role for people in nature, but "a role for nature amid humans. Most wilderness advocates were urban dwellers who thought of wilderness as part of an urbanized society." Cronon's analysis, he continued, "reflects the temptation for historians to draw into their historical analyses both personal and moral struggles and the ideology of contemporary debate." This, he concluded, leads to bad scholarship and obscures actual opportunities in environmental history.[46]

Others went beyond Cronon's approach and what they saw as the political implications of his view to the foundations of his ideas.[47] *Wild Earth* devoted most of an issue to the book that came from Cronon's seminar, *Uncommon Ground,* and most of that to "The Trouble with Wilderness." The cover read: "Opposing Wilderness Deconstruction: Gary Snyder, Dave Foreman, Don Waller and others respond to the latest attacks on wilderness," and Foreman began his editor's column by declaring that "this issue of *Wild Earth* casts an eye to the dirt clod Professor William Cronon recently tossed at the Wilderness Act and at defenders of Wilderness Areas." Although he did not think Cronon was hostile to nature, he believed that his "complaints are based in ignorance of biology, a misunderstanding of the conservation movement, and a carelessness about the consequences of his critique of wilderness."[48]

All the articles but one took some variation of this approach, uniting in the belief that wilderness is real, outside human society, and beyond human control, though the arguments were not always to the point. "If," Donald Waller said, "wilderness is, admittedly, a very human construct laden with cultural meaning, wildness is just the opposite: that which is not, and cannot be, a human construct," thus agreeing with Cronon in the first half of the sentence, and in the second rejecting a position that Cronon did not take.[49] Gary Snyder, whose piece ran under the head "Nature as seen from Kitkitdizze is no 'Social Construction,'" wanted to "take these dubious professors [who see nature as a social construction] out for a walk, show them a bit of the passing ecosystem show, and maybe get them to help clean up a creek."[50] Bill Willers, in "The Trouble with Bill Cronon," said that "Cronon may be correct that *ideas* of nature don't exist outside of cultural understanding, but Nature in all of its self-governing complexity most certainly does."[51] Another piece quoted Donald Worster's comment that "one of humankind's oldest intuitions is that the realm of nature has an objective, independent order and coherence; that we are to some extent part of that order . . . that in any, [sic] case we ought to respect it."[52]

Most of the critics had attacked a case Cronon had not made. They took his statement that concepts were social constructions to mean he believed what was "out there" was entirely a social construction, which he did not. They said he confused reality and text, but he argued—paradoxically and somewhat ironically—that this was what wilderness's defenders too often did. Seeing what was out there through the lens of "wilderness," they mistook the name for the reality. Although the critics accused Cronon of the academic and political sins of postmodernism, naiveté, and relativism, their attack came less from academic debates than a fundamentalist understanding of wilderness. Like fundamentalist Christians refusing to see the Bible as a collection of texts produced at different times and within particular cultures, they put wilderness outside historical and cultural context, believing it could not be analyzed but had to be experienced. If we substitute "wilderness" for "Bible," their positions looked much like those of Biblical literalists, who believe the Bible speaks to all who come with an open heart and that the Bible is the only source of light.

If directly challenged, doubtless all Cronon's critics would have admitted that ideas about wilderness were not the same thing as the wilderness and that culture affected what we think. Some, in fact, made these points in their critiques. They acknowledged the complexity of the issue (as an intellectual problem) and did not quite appeal to the insight of the unlettered, which confounds the learning of the wise. Still, they saw wilderness as ultimate reality, directly accessible to the believer, which stopped analysis. If the value of wilderness lay in showing humans ultimate reality in moments like Thoreau's on Ktaadn, then academic arguments became at best irrelevant, at worst impious mental meddling with something beyond our ken.

Besides the reaction to Cronon, the stated purpose and the rules for wilderness travel showed, less passionately but more clearly, how deep the belief in sacred wilderness ran. J. Baird Callicott, environmental philosopher and Aldo Leopold's interpreter, said that "one of the principal psycho-spiritual benefits of wilderness experience is said to be contact with the radical 'other' and wilderness preservation the letting be of the nonhuman other in its full otherness."[53] That did not put the case strongly enough. Contact with the "radical 'other'" lay at the center of wilderness's benefits, and by the 1970s trips into the wilderness acquired the quality of a pilgrimage or quest. Earlier values, like the chance to test one's character against ordinary difficulties, the reenactment of pioneer life, or the delights of self-sufficiency, faded, with the physical difficulties forming the setting for spiritual struggle. That mixture had fewer contradictions than the formula suggests. Most spiritual traditions saw the body as a way to the spirit and physical difficulties as the path to insight, and while wilderness travel as pioneer recreation had strong overtones of conquest and emphasized competition, the tradition also believed contact with nature made for virtue, and the ethos of outdoor recreations like mountain climbing and hiking emphasized the conquest of physical nature as a way to self-knowledge and spiritual insight.

The wilderness movement, in a stroke of genius or insight, harnessed pioneer activities and outdoor recreations to a spiritual search, using the familiar appeals of pioneer nostalgia and the strenuous life for less familiar ends. It emphasized immersion in the land and actual effort. By fully entering the world of nature, we would pass beyond our usual limits and would find a personal, emotional understanding of our situation. Wilderness's defenders built on Emersonian intuitions, but where earlier advocates of finding insight among nature associated spiritual experience with dramatic scenery or comfortable settings, wilderness looked for the fully wild and believed in discomfort. Instead of going to majestic groves of trees, where they could stroll without being hindered by undergrowth, or sitting by mountain lakes or in alpine meadows, looking up at snow-capped peaks, pilgrims canoed down Arctic rivers or hiked down desert canyons. Instead of vistas they experienced intact ecosystems; rather than aesthetic delight, full sensory experience. Few spiritual traditions recommended sweat and blisters as the way to the mysterium tremendum, the "radical 'other'" experienced in all its beauty and terror, but wilderness sent the seeker off for enlightenment while wet and cold or hot and tired, suffering thirst and a sore ankle.

The journey into intact nature became a sacred quest by virtue of the taboos that hedged it off from recreation or vacation, conventions that guided the pilgrim to an experience of the wild on its terms. The taboos relied on a common intuition about the sacred, that it was not defined but encountered, and that ritual guided the encounter by helping the seeker to step outside daily life. By the 1980s custom made the essentials of the wilderness journey as spiritual search as stereotyped as the actions required for a plenary indulgence. Accounts of wilderness travel in books and magazines, even in the travel sections of newspapers, began with leaving modern transport. If people chartered a small plane to get to a remote wilderness lake, for instance, their experience only began when the sound of the plane's engine faded in the distance. Wilderness travel did not demand primitivism, though. It banned technology only when, as Aldo Leopold said of gadgets for hunting, it passed the "limit beyond which money-bought aids to sport destroy the cultural value of sport."[54] Where technology preserved the wilderness, proper wilderness traved demanded its use. Wilderness travelers did not sleep on pine boughs they cut each night or cook over open fires because even small numbers of people using wood for campfires and boughs for camp beds would destroy popular areas. They had to use stoves and matress pads. Arguments on other matters grew heated. Cell phones had no place, but some believed they could be used in emergencies. People discussed the conditions under which such ordinarily banned technologies as helicopters might be used, say for rescuing sick or injured travelers, or how many people could be in a wilderness.[55] Purists grumbled about even seeing other parties at a distance, the less pure ignored or tolerated long-range reminders of other humans. How wild did wilderness have to be? One school of thought emphasized size; another, visible lack of human change; still another, an ecosystem intact to the top of the food chain, where there was the chance, however slim, of participating in the food chain below the top. Not even the fanatics, though, insisted you had to be eaten by the grizzly to get the full wilderness experience.

All these debates, which continue in publications like *Outside* and *High Country News,* rested on the premise that the holy had to be approached and experienced in a way consistent with its holiness. Wilderness seekers did without ATVs for the same reason people engaged in Zen meditation turned off their cell phones. Spiritual search required a break with daily conventions in favor of patterns of behavior that fostered contact with what lay beyond ordinary life.

The idea of wilderness peace, if not enlightenment, proved so attractive that the commercial drives Leopold lamented in hunting seized on images of wilderness as avidly as a child grabbing cookies after school. Advertisements for off-road vehicles presented them as magic carpets. The right SUV took you from the traffic jam directly to a mountain road with no traffic, and when you turned off the road it unrolled a highway beneath your wheels that led to a mountain meadow or the edge of a canyon. Motorbikes existed, in the ads, to take young boys and their fathers to fishing holes, where they would catch fish and form family bonds. Corporate workshops sprang up, promising that three days in the wilderness would pull together a management team so effectively that they would sell 40 percent more soap. Gurus ran weekend wilderness "workshops" that brought you into communion with the earth. Ecotourism offered wild equivalents of guided tours of the Holy Land, on every level from the tour with his Grace the Bishop and a personal audience with His Holiness on the return trip down to the charter flight with Father O'Malley that threw in a ticket to the Vatican museum. You got the wilderness effect by turning your back on the *Princess Squaresides,* which had brought you to the penguin rookery.

The Wilderness Act of 1964 established wilderness as a land use but did not decide the issue (this one, unlike the first, debated as much within the wilderness movement as outside it) of the place of wilderness in American life and its value to modern industrial society. Here the arguments and programs the Wilderness Society's founders used did not serve. They extolled wilderness as a place to practice woodcraft and test manhood against the wild. The pioneers' appeal was fading, though, and mimicking pioneer travel did not, in any case, connect people's lives to the wild.

Sacred wilderness drew on new ideas and presented new arguments while retaining the attractions of the old. Against the abstractions of argument and the alienation of modern life wilderness engaged people through their senses. Approaching sacred wilderness required walking the land, savoring the scents of pine or sage or rock dust, feeling grit or soil under their boots as they balanced for a next step, listening to birds or the wind. It brought wilderness to people and also gave flesh to science's theories of intact ecosystems. Sacred wilderness, though, used old perspectives and had its own limits. It held to nature appreciation's focus on the individual and on experience in nature as only a part of life rather than addressing directly ecology's central claim that we live immersed in nature, that we belonged to what Leopold called the community of life not by choice but by necessity. Environmentalism, though, asked for more than personal connections and more than sensations gathered on vacation. Our numbers and technology, it held, made human engagement with the world the central problem of our age.

That belief, which marked the change from nature appreciation to environmentalism, forced a question that wilderness had not yet addressed: the values of wilderness not for the individual but the society. Out of the wilderness the prophets came. What signs did they have and what messages did they bring? People do not venerate the Buddha for becoming enlightened, but for returning to the suffering world with the message of enlightenment. Christian mysticism insisted that the monastic life, even the hermit life, had a social end. The monk might seek the desert to purge his sins, but his end was to reshape the "world," if only by his prayers. The prophets of the environment called us to repentance, for the Kingdom of Nature was at hand, but how were we to dwell in that kingdom? With that question we come to the heart of the environmental passion. Beyond rational and efficient resource policies and the beauties or complexities of nature, the movement insisted that our species was at a crossroads, that we were destroying the basis of our lives and that our salvation could come only if we forged a new relationship with the world, one that would heal it and nourish us. Environmentalism emerged as people accepted that point and turned to shaping their lives and their society.

The rest of this book asks how people, over the last forty years, have changed their values and their lives in response to their new understanding of their place in the world, and that requires a shift of focus. Environmentalism's heritage in Western culture and American history involves high ideas, high ideals, and broad cultural currents, while its development calls for ordinary action in daily life. Environmentalists used ecology's descriptions of our species' relationship with nature to turn abstract and distant ideas into concrete, individual moral duties and to guide their search for the ultimate, but they had no road maps. They had to learn from experience. These last forty years divide, not always neatly, into two parts, the years of crisis, roughly the late 1960s to the mid-1980s, and the movement beyond crisis to concern with the long-term, which began in the 1980s and continues today. The movement developed during a time of great hopes and fears, when people expected the millennium or the collapse of industrial civilization. As it became clear that the industrial system would not collapse tomorrow and that sustainable society was not around the corner, people became more concerned with living in a society while changing it and with the implications of environmental ideas for action over a lifetime. With that last we come to the present, for we now face the questions of how to build a culture and a society that will allow satisfying lives and preserve earth's life. We also come back to the topic we started with, environmentalism as an expression of the religious impulse. If environmentalism is involved with ultimate questions, if it speaks not just of the beauties of nature, human health, or a sustainable economy but to our relation with the universe, how should environmentalists understand their movement? That we will take up in the conclusion.

4 / Sacred Nature Enters Daily Life

Environmentalism emerged as a movement when people applied an ecological perspective to their lives and society, seeing the world as webs of relationships rather than separate things. Environmentalism did not find its name until the late 1960s, but people's ideas began to change in the years after World War II; formal and informal nature education made them familiar with ecology, and events made them aware of their involvement with nature. Elementary school nature stories and lessons helped, showing creatures in relation to the land. So did things like the Boy Scout handbook, which described the process of plant succession. The Scouts' parents found out about their connections to the world through news stories describing the jet stream, which brought radioactive isotopes from H-bomb tests in the middle of the Pacific Ocean to America, and bioconcentration, which accumulated the substances in food. American mothers learned that the milk they gave their growing children "to build strong teeth and bones," as the advertisements so cheerily put it, contained strontium 90 that might settle in those bones and cause cancer (this during the decade when leukemia seemed the childhood plague).

In 1957, when federal inspectors announced, a few weeks before Thanksgiving, that they had found some cranberries contaminated with a pesticide that caused cancer in laboratory animals, it caused a national, if short-lived, panic. Five years later, *Silent Spring* crystallized public fears about environmental problems, exposing the dangers from pesticide residues in body fat and even human milk, and the possible, though unknown, risks of cancer, deformities, and mutations. A steady diet of bad news kept the issue alive. In Cleveland, chemicals in the Cuyahoga River were so concentrated that the river caught fire (it had in the 1950s as well, but no one had paid attention).[1] Scientists predicted that industrial wastes and chemicals would kill all life in Lake Erie. An oil well in the Santa Barbara channel blew out, coating that city's beaches with crude oil and the evening news with pictures.

In the 1960s the level of public awareness changed rapidly. When *Silent Spring* appeared, only scientists were using the word "environment." Eight years later, Congress enthusiastically passed the National Environmental Policy Act; the federal government established the Environmental Protection Agency and the President's Council for Environmental Quality; and a quarter of a million people rallied in Washington, D.C., for the first Earth Day, where they heard scientists and senators speak about the urgent need to save "the environment." President Richard Nixon, who had until then successfully concealed his interest in the subject, proclaimed 1970 the year of the environment and called for new and more stringent laws to protect it. New organizations sprang up to advance an "environmental" agenda, while established nature protection groups changed their rhetoric and programs. Environmentalism became a worldwide cause. By 1970 every noncommunist industrialized nation had its environmental lobby, and in Australia, New Zealand, and Europe activists formed "green" political parties.

Threats to their health and their children's health drew people, but a climate of change encouraged them as well. The oldest baby boomers, in their college years, challenged authority and marched for social justice and legalized marijuana, while race riots, dissent from the Vietnam War, and the counterculture gave force to visions of a new society. Environmentalism began in an atmosphere of hope and fear, amid suspicions about authority and science on one hand and enormous confidence in ecology and (sometimes) technology on the other. Reformers rejected American society but believed in individualism and consumer choice, pointed toward a millennium of peace and feared the destruction of all human societies, fostered utopian schemes but demanded neighborhood action. Every street corner had its prophet, and there was much talk about "the revolution," an event that varied in form and cause from speaker to speaker but always involved dramatic social change.

Scientists' warnings about the imminent destruction of the earth's systems of life fit both apocalyptic and millennial visions. Pessimists believed that pollution and resource depletion would destroy nature's balance, and a few believed the crash would wipe out the human race, leaving rats and cockroaches to inherit the earth. Optimists hoped a new society would rise, phoenixlike, from the ashes of the Establishment, using ecological insights to cure the alienation—of people from themselves, from each other, and from nature—that was a consequence of industrial civilization. With varying degrees of enthusiasm, knowledge, and wisdom, people tried to put their principles into practice. Those who were deeply committed organized new communities on new principles; the rest purchased copies of *Ecotopia,* a utopian novel that depicted a new, environmentally sensitive culture.[2]

Beneath hope of change lay the horror of losing the world of wild nature, and environmentalism moved from obvious problems to this deeper level. Rallies against chemicals like DDT led to bans on some, but that did not change the conditions that caused farmers to spread pesticides on their fields. Protests about automobiles brought pollution controls and mileage standards, but these failed to reach the economic and social conditions that encouraged suburbs, malls, and parking lots. Americans struggled with what it meant to be connected to the world as ecology said they were, and how they, as individuals and together, might put principles into practice. Environmentalism believed the task of forming a right relation between our species and the world was the crucial problem of the age, one that involved every part of our lives and every part of our society. It could rely on nature appreciation and the wilderness movement for a spirituality of nature, but it had no help in its new task: building a morality of nature. People approached this from two directions: green consumerism tried to change society from within, while bioregionalism looked for a new kind of daily life outside the established economy and in tune with the land.

DAYS OF HOPE AND FEAR

Great hopes and awful fears colored environmentalism's early years, but optimism prevailed. This was, after all, America. Ernest Callenbach's *Ecotopia,* a product of the counterculture's embrace of environmentalism, proved the most popular expression of that confidence. Printed in 1974, it quickly went from specialty press publication to mass-market paperback and on to commercial success. *Ecotopia* did not owe its success to its literary qualities. The book had to be called a novel because that was the conventional term for long fiction narratives, but the writing was poor, the characters cardboard, and the plot only an excuse for the ideas. The ideas themselves did not wear well. Twenty years later its domestic technology and electronics seemed quaint and the social ideals too obviously the enthusiasms of the 1960s. The book remained in print, though, because its vision looked beyond the mood and hopes of a generation to the deeper and continuing desire for a society and a life that made sense.

Set in 2000, *Ecotopia* described the ecologically responsible society of Ecotopia, formed in 1976 by the secession of Washington, Oregon, and northern California from an unreformed and polluted United States, through the eyes of one of the first American visitors. Admitted in 2000, the newspaper reporter William Weston found an almost perfect society organized around the care of the earth. Ecotopians, guided by ecology and psychology, lived in harmony with themselves, each other, and the world around them. Education stressed their connections to the world, and life reinforced those lessons. When building a house, for instance, they cut the trees themselves or did other work that allowed them to see where their homes actually came from. Public policy, having already deeply cut pollution and reduced energy use, aimed at a steady-state society in which the human population took from the land only what its ecosystems provided.

Ecotopians used science to live with each other as well as with nature. Social mechanisms that ranged from communal families through work schedules to ritual (if not deadly) combat eliminated or gave sanctioned outlet to undesirable human traits like aggression. Making peace with the earth and each other allowed Ecotopians to make peace within. The society encouraged self-development and self-realization through face-to-face contacts and a maximum of individual freedom. True to the utopian novel's conventions, over the course of the story Weston's reactions progressed from suspicion to acceptance to enthusiasm, and in the end he turned from American society toward the environmental future.

Ecotopia mixed many things. In literary terms, the book broke no new ground. It used the apparatus of the utopian novel—the isolated land, the perfect society, the innocent narrator from abroad, the citizens ready to spout statistics and philosophy at the drop of a conversational hat—and that genre's key premise: that a society organized around a correct understanding of human nature produced happiness and peace. By setting that life in nature, *Ecotopia* followed pastoral and agrarian utopians. Where it broke with tradition was its mixing a modern society with nature; its happy citizens lived in cities and learned science, but hunted and celebrate a successful hunt with dancing. Ecotopians lived in a peaceful society and lived in the moment, but had challenges and opportunities for growth. They worked together, but enjoyed individual liberty and the chance to shape their government.

However, *Ecotopia* clung to old values and established ideas—more than most of its readers probably realized. Although it allowed plural marriage and barred private cars, frowned on getting rich, and spoke against Wall Street's economic values, the revolutionaries held as firmly as the stockbrokers to individual autonomy; Ecotopia depended on science as much as the military-industrial complex; and Ecotopians embraced technology with the fervor of a lifetime reader of *Popular Mechanics*. Ecotopia's pure democracy and open government depended on television; its sustainable society, on ecology and the biological sciences; and its social arrangements, on psychology. The novel held that knowledge that in the old society had alienated people from each other and given power to the ruling elite would in the new provide the foundations for self-realization and democracy.

Currents of fear ran almost as strongly as currents of hope. To many it seemed our civilization had already gone too far and rather than harmony with nature our future held only a struggle for life amid the ruins of nature and those of civilization. Science fiction contributed its share of warnings (aficionados of gloom will recall John Brunner's *Sheep Look Up*) and science even more. Barry Commoner's *Closing Circle* (1971), a warning about dwindling resources and collapsing ecosystems, had two hardback printings in a year, three book club adoptions, and five paperback printings in three years. Garrett Hardin's article "The Tragedy of the Commons" (1968), which argued for regulation of common resources (as open use led only to disaster), set off a long public debate, and its title added a phrase to the language. Paul Ehrlich's *Population Bomb* (1968), which pointed out the long-range consequences of the explosive population growth after World War II, had the same effects. The greatest uproar accompanied the Club of Rome's report on the world situation, *The Limits to Growth* (1972).[3] Its suggestion that the world faced environmental and social disaster in a very few centuries unless economic growth ended drew condemnation from around the world, from Mao Tse-tung to the free-market economists of the University of Chicago.

People spoke of these works as jeremiads, and they were apocalyptic literature, in the classical sense of the term. They did not predict doom but called people to repentance. Put in conventionally religious terms, these works pointed to the hell that would result from our sins, called us to change our ways, and in some cases described the new heaven before us. Donella Meadows, who worked on the Club of Rome study, protested that *The Limits to Growth* had not been about prediction, but choice. The study had set out to "challenge the myth of growth as the answer to all problems" and had called us to another path.[4] Commoner began *The Closing Circle* with a discussion of our reluctance to admit that the world is fully connected and that we are part of it, but he believed we could change and ended with a call to action. Ehrlich, for all his statistics and warnings about an explosive increase in people in *The Population Bomb*, believed we could curb population growth. Indeed, he put much of his energy into arguing that the sooner we began to do that, the less pain the human race would suffer. Zero Population Growth, formed to address the population issue, called on people to change their behavior, not abandon hope. Even the anti-environmentalists a generation later conceded the point in a backhanded way. In *The Skeptial Environmentalist*, Bjorn Lomborg triumphantly pointed out that the "predictions" had not come true because humans had changed their ways.[5]

Environmental enthusiasm produced studies, laws, and arguments, but it also reshaped the political and social landscape of nature protection. Into the late 1960s groups formed around 1900 or in the Depression years dominated the scene, and they divided nature into special interests. The Audubon Society tried to save birds; the National Wildlife Federation, wild animals; the Izaak Walton Leagues, fish and the clean water they needed. Each used conventional and usually genteel methods—mainly nature education materials in the early grades, public policy statements, and lobbying. They rarely engaged in public battles, like the ones over Echo Park, and then only reluctantly. The environmental cause drew people in numbers that swamped the old organizations.[6]

By the early 1970s, the National Wildlife Federation, originally a coalition of hunters, fishermen, outdoor enthusiasts, and conservation officers, included every shade of opinion, from sport hunters to advocates of humane treatment of animals who wanted to abolish hunting. New ideas and new programs produced upheavals from above and below. Radicals accused leaders of lacking the proper concern for the environment, while others reined in leaders they saw as too aggressive or not focused on the group's established priorities.[7] David Brower threw himself and the Sierra Club into the environmental fray, alienating the respectable, upper-middle class professionals who formed the club's core even as he attracted a younger and more radical group that would form a new core. In 1969 the Sierra Club board removed Brower as executive director. The sense of urgency that came with environmentalism made lawsuits, once emergency measures, popular (at least until costs became too high) and brought new groups to power. The Environmental Defense Fund began in 1967 as a private group with ten members, a legal arm consisting of one of the founders, Victor Yannacone, and a scientific staff of a few academics. A decade later, it had a general membership, an office in Washington, D.C., and full-time staffers who administered programs and lunched with assistant undersecretaries in the Environmental Protection Agency.

Below the visible level of laws and lobbies, environmentalism's ecological view—of nature as systems rather than parts—set off a debate that pitted environmentalists against defenders of the conventional economic wisdom and divided the movement itself, for the new values ran against some ideas entrenched in nature protection. Those who defended wildlife from humane convictions found themselves, for instance, at odds with others who emphasized ecosystems over organisms. Introduced herbivores— feral burros in the Grand Canyon, mountain goats in the Olympic Mountains—threatened native flora. Those most concerned about the pain or death suffered by individual animals wanted to capture them alive; those concerned primarily with the ecosystem accepted or even favored shooting. The description of the union of animal liberation and environmental ethics as a bad marriage followed by a quick divorce caught the tone.[8] That debate, though, evolved as humane advocates, recognizing that the intellectual foundations of their position rested on nineteenth-century biology and psychology and a program developed to save domestic animals from cruel treatment, asked what their ideals meant when applied to creatures living in natural systems seen in ecological terms.[9]

Besides disagreeing over how to solve environmental problems, environmentalists were divided over their cause. Some blamed runaway technology, others, science or the wrong kind of science. One faction believed the population was central, others that the ideology of conquest at the base of Western civilization was the key. Marxists, anarchists, and countercultural gurus incorporated environmentalism or rejected it as a distraction or divided on what the movement meant in their ideology. Academic cottage industries sprang up to guide technology into the proper channels, develop a nonhierarchical science or a science that incorporated feminist perspectives, make Marx green, or use anarchism as the foundation of a new, locally based society with appropriate technology.

Although environmentalists discussed the implications of their views, action more than ideas helped them. People wanted to "make a difference," and they refused to wait for analysis to guide them. They acted, and in action, more than in reflection, found or forged their principles. Because everyone could lend a hand, saving energy or resources became a popular crusade. Articles and letters to the editor began with the phrase "If we all . . . ," then continued through elaborate calculations of what would be saved if everyone bicycled to work, recycled glass bottles, or added three inches of insulation to the attic, and ended with a description of a new humanity caring for the planet.

Debates about cloth versus disposable diapers and permanent-press versus all-cotton shirts produced calculations and logical arguments as serious and elaborate as the legendary ones among medieval theologians about the number of angels that could dance on the head of a pin. People formed and re-formed local coalitions to save energy or recycle newspapers in their neighborhoods. "Reduce, reuse, recycle" became a mantra. *Sierra, Audubon,* and *National Wildlife* sprouted sections—some on the coarse recycled paper then available—telling people how to live simply and have less impact on the land. The *Whole Earth Catalog,* that middle-class guide to countercultural consumption, reviewed and recommended books and gadgets and technologies for the environmentally committed.[10]

Environmentalists gave practical reasons for saving energy and resources, but they looked beyond dollars and BTUs for a new relationship between people and the earth, one that came from placing our conduct toward nature "in the realm of morals" and in the context of our obligations to life on earth. That perspective made energy-efficient appliances a moral matter, recycling a question of virtue, and environmental philosophy into spirituality or practical theology. Most of us think of "moral practice" in terms of fasting and prayer or preachers condemning liberal lifestyles and associate "spirituality" with Great-Aunt Rose's candles, Eastern mystics, or self-help books promising to increase your income 20 percent if you pray for five minutes a day—in any event with actions and ideas outside ordinary life. "Theology" conjures up images of the Inquisition or of old men parsing the fine points of obscure systems of thought. In religion or religious studies, though, moral practice might involve fasting and prayer, but, more generally, it refers to daily action that expresses or strengthens faith. "Spirituality," in a modern definition, means "an awakened consciousness of the sacred, and an ordering of life toward that consciousness."[11] Theology, technically, is "rational inquiry about religious questions [or an] organized, often formalized body of opinions concerning God and man's relationship to God." It is the logical articulation of beliefs in the context of experience—in an older definition, "faith seeking understanding."[12]

Applying those definitions to environmentalism turns daily life into a series of moral choices about what kind of shirts to buy and car to drive, or whether to drive at all, for these decisions affect the world's future. Environmentalists' practices look very much like what Christian spirituality referred to as the "conversion of life," the state in which we examine how even our smallest actions and decisions orient us toward, or away from, the path of right living, for environmental counsels about responsible consumerism are as detailed as religious guides to moral living.[13] "Spirituality" seems the only appropriate word for exhortations about identifying with the living earth. "Theology" is equally apt for the systems environmentalists used to explain the world and justify their beliefs. Environmentalism's aim of transforming ordinary life in order to save the planet moved beyond reform and the individual spirituality of nature appreciation and the wilderness movement.

Environmentalism's concern went beyond spirituality or individual development to the level of religion, for it required, beyond changes in individuals' lives, a shift in society's values. It did this, though, without appealing to supernatural elements. Spirituality it accepted, transcendent spirits it did not. Environmentalism did not claim humans had, individually or as a species, a conscious relationship with the forces of the world, but it believed we had an obligation to act by the laws of the universe. It rejected the conventional perspective of a world made up of pieces, in which human actions had only limited consequences—the idea that we could choose what streams to pollute, which to reserve for trout fishing, set aside some areas as national parks, and then let Progress in all its forms run right up to the fence. Events challenged that view, and by the late 1960s, many Americans were searching for ways to live as citizens of the biological community, and they believed we all must learn to live that way.

Environmentalism made far more concrete claims about human connections to nature than the Romantic understanding that everything was hitched to everything else in the universe—that we could not move a stone without the troubling of a star. The connections it made between humans and nature more closely resembled abolitionist arguments that linked free-state prosperity to slave-state cruelty, using economics to tie New Englanders' daily comforts to cotton grown under the lash and the fat farms of the Midwest to the slave markets of New Orleans. Environmentalism applied ecology in the same fashion, tracing the destruction of the rainforest to our appetite for inexpensive fast-food hamburgers and connecting dried-up riverbeds to our demand for lush lawns and sparkling cars. Like abolition, environmentalism went on to argue that we had a moral duty to act and that small changes in our daily actions could have enormous consequences for the world, and it required individual action on a fundamental level. Radical abolitionists had demanded that Americans reject a system that allowed and supported slavery, including the Constitution, which William Lloyd Garrison famously called a "league with Death and an agreement with Hell."[14] Environmentalists demanded something similar, though on more than a political level. They wanted us to move beyond better forest management, more wilderness and national parks, and sensible pollution laws. They asked us to change our ways and our lives with the aim of ultimately changing our society.

Environmentalism's concern for a responsible life attracted a few philosophers who wanted to lay its foundations, or locate them in the Western intellectual tradition, and a larger number of radicals who wanted to integrate its insights with their own systems, but before they could be described, principles had to be developed, and that required trial and error.[15] Environmentalists followed two paths. Some tried to live responsibly in the existing society, others to make a new life outside industrial America. The first path satisfied those who either saw no need for radical action or no immediate way to it. They recognized that recycling, conserving energy, and living might not be enough to save nature, but these actions helped. Millions made at least temporary changes in the way they lived, and many continued to think about what they did and why they should do it even after the first wave of enthusiasm ebbed. Others, though, believed that working within the established order was either impossible or useless, and they set out to make a new life that would restore nature's communities and build a social order others could imitate.[16] Green consumerism and bioregionalism gave people the chance to put convictions into practice, but, more important, allowed them to make sense of practice by finding out what principles and moral lessons meant in the context of life.

A NEW LIFE

For many, a new environmental life began with the understanding that we all needed to change the way we lived. People took up many causes, but they made energy conservation—with the help of the oil crises of 1973 and 1979—a national program. Taking advantage of federal tax credits for alternative energy systems, they hired hippie contractors to put in solar systems for heating and hot water. They put more insulation in the attic and walls, bought new storm windows, plugged cracks around baseboards and light switches, and wrapped insulation around the hot water pipes in the basement. A few tried to "get off the grid," declaring freedom from the power company, but most simply supplemented conventional systems. Early efforts required some patience and commitment. Homeowners had to open and close the shutters on solar greenhouses every day and, when the builder went out of business (which happened with depressing regularity), do their own repair and maintenance work, calling around town, for instance, to find propylene glycol to pump up the solar hot water system.

Most of the plans people pushed with such fervor proved more trouble than they were worth, but some did not, and by the late 1970s established firms replaced the hippie contractors, and factory-built systems designed by engineers took the place of jury-rigged pipes and collectors. Home windmills and generators proved inefficient, but on a large scale wind energy turned out to be profitable. In the 1990s engineers designed enormous pylons, mounting giant blades hooked to generators, and wind farms sprang up in parts of the country with high, constant winds.

At the other end of the spectrum, nature became an upscale consumer item. The Nature Store, for example, offered environmentally friendly nature goods—nuts from the rainforest gathered by indigenous peoples, games to teach children the principles of ecology, and environmentally correct books (animals had to be animals—no bunny rabbits in overalls allowed).[17] Located in high-end malls, suffused with an air of authenticity and caring, they nevertheless operated with an efficiency surpassing the Chicago meat packers, whose legendary boast was that they used all the pig except the squeal. The Nature Store sold recordings of the cries of birds and whales and the sounds of forest and ocean in forms suitable for the high-tech sound system of an apartment or suburban home.[18] Environmental education became part of middle-class childhood socialization, with books on environmental action, games, and decks of playing cards featuring endangered species. Elementary schools added environmental lessons, and some three dozen books, including a set of school texts in English and Spanish, had the title or subtitle "How to Save the Planet."

Far from flagging after the energy crises, getting right with nature became more popular. Green guides appeared to help "you make environmentally responsible choices in everyday decisions, and enable you to audit your personal impact on the earth," and *E/The Environmental Magazine* showed you how "your own actions and those of others affect our entire ecosystem."[19] The Sierra Club's *Green Guide* had 200 pages listing firms specializing in or offering goods and services from diapers and washcloths to retirement funds that invested only in environmentally responsible corporations. Businesses advertised their energy savings or boasted about their recycling.

The plastic packaging of Green Forest paper towels and toilet paper sported outlines of evergreen trees, the percentage of post-consumer content, and the motto "The right thing to do for the environment."[20] Environmental magazines exhorted readers to turn off unneeded lights, put the newspaper in the recycling bin, and buy windows with better insulation, and pointed to their own example. The *Amicus Journal,* bulletin of the National Resources Defense Council (now *One-Earth*), informed readers that it was "printed with vegetable-based inks on 100 percent recycled paper that is processed without chlorine," and it offered a "Living Green" column to help consumers put their environmental convictions to work in their homes and daily lives.[21]

Economics accounted for some of this enthusiasm, hypocrisy for some more, but there was plenty of genuine moral passion, commitment, and a strong sense of evangelism. As the *Consumer's Guide to Effective Environmental Choices* put it, when we take the environment into account in what we buy and use, we not only "send important messages to manufacturers . . . we let our family, friends, and neighborhood know something about our values. . . . [Public] actions can have an impact. . . . When you do something that your friends and neighbors can see, you go on record as being concerned about the environment and you act as a role model for them." The guide assured people that they did not have to be community leaders: "In retrospect, the triumph of recycling is especially impressive because so much of the change in attitudes and individual behavior was instigated by seemingly powerless children and teenagers who prodded their families, schools, and colleges into action." It applauded *Sesame Street*'s campaign for recycling, adding that "probably most parents at one time or another have received humbling lectures from their children after being caught tossing a glass jar or newspaper in the trash can."[22] And a little child (or possibly a teenager) shall lead them.

Readers raised in pious households will recognize these sentiments, perhaps recall lectures about the need to do the right thing and the satisfaction that would, eventually, come from it, even if your friends laughed at you now. Pre–Vatican II Catholics can hark back to those campaigns to save pagan babies by donating five dollars to the missions; Jewish children, to pushkas and shalakmones at Purim; American Muslims, to small gifts of food offered to puzzled suburban neighbors at the end of Ramadan. As with these explicitly religious cases, actions meant less than moral lessons and demonstrations of faith. Anyone who could count above ten without taking their shoes off knew that individual efforts or even organized campaigns would not save enough paper, plastic, oil, wood, or water to make a dent in the load America's energy-hungry society imposed on the world's ecosystems. What it could make was good habits in children and good examples in the neighborhood.

While many found it enough to recycle their trash and use canvas grocery bags, others feared those actions gave only a false sense of security and the illusion of virtue. Believing we had to make immediate, radical changes, they looked for new ways of living that would heal them and the land and serve as an example to others. Like green consumerism, bioregionalism had a quite Protestant emphasis on individual action and individual salvation, to which it added a view (also Protestant) of existing social arrangements as unregenerate and possibly damned and the distinctly

American tactic of making a new community that would be as a city set upon a hill.

The tradition of an "errand into the wilderness," as the Puritans called it, stretched back to seventeenth-century New England and peaked during the revivals that swept the Northeast between 1800 and 1840, the movement historians call the Second Great Awakening. That burst of enthusiasm included the Oneida and Amana communities, bywords in those years for their open sexuality; the Shakers, known for celibacy; and the most durable of American religious settlements, the Church of Jesus Christ of Latter-day Saints, its impress still evident in the Mormon empire of Utah and the intermountain West.

Secular utopias dated back almost as far. Several of the American colonies had begun as planned societies; in the early nineteenth century Robert Owen founded New Harmony, Indiana, established on Fourierist principles. Every generation after industrialization had its enthusiasm for going "back to the land," breaking with the wage-labor slavery of factory and office and living independently on a small plot. The counterculture's communes and explicitly environmental utopias may have seemed radical, but they stood in a long line of American experiments.

Like their predecessors, most of the environmental experiments quickly collapsed, but enough—and enough enthusiasm—survived to fuel a movement. Enthusiasm was necessary, for bioregionalists had little to guide them. Ecology's theories were good, but concepts of community and the laws governing flow of energy through an ecosystem gave about as much direction to bioregionalists trying to farm responsibly as the laws of thermodynamics did to an engineer building a steam engine. People needed, besides theory, practical knowledge. Bioregionalists, in fact, needed very local knowledge. Irrigated farming with stock as an adjunct to crops required a different way of living with the land and with the neighbors than subsistence farming in a well-watered country, and soils varied, along with weather and plants. In many ways, at different speeds, and with different degrees of success people found the intimate, practical knowledge they needed to make a living.

Social knowledge proved as essential as biological knowledge, for bioregionalists wanted not just to live on the land but to live in a community. They had even less guidance in dealing with human neighbors than their animal ones, for there was no science of human community, and the ways of the people living around them usually did not aim at the ecological health of the land or include nonhumans in any way bioregionalism would accept.[23] Primary cultures offered examples, but not ones bioregionalists could easily use, for earlier societies had different tools and worked under economic and social conditions very unlike the edge of industrial America. Bioregionalists persevered, though, and by the early 1980s had enough of a movement for *CoEvolution Quarterly* to devote an issue to it and for the *Whole Earth Catalog* to add a bioregional section.[24]

Although bioregionalists often talked like back-to-the-land enthusiasts, they had much broader aims. The people with a copy of *Ten Acres and Independence* in their back pocket wanted to live on the fringes of industrial America; bioregionalists looked for a way to live that would change industrial America. The *Whole Earth Catalog* described the movement in political terms, as a reaction to a distant federal government, a return to self-rule and local involvement, but also "a recent revisioning of North America" that emphasized a regional culture and local and immediate experience, self-reliance, and ties to particular landscapes. Gary Snyder spoke of bioregionalism as an instrument for social change, not "just a rural program [but also] for the restoration of urban neighborhood life and the greening of the cities."[25]

That caught the radical core. However much bioregionalists sought to live their own lives, they were less interested in escaping society than transforming it. They did not seek a niche for themselves or to return to the economy of independent producers that had dominated the early republic. They insisted, as no earlier agrarian reformers had, on including the land in the community, on giving it a place and a claim on humans. Agrarian reformers sent people back to the land to live from the land; bioregionalism sent them to live with the land and nurture it. Agrarianism regarded humans as stewards but measured their obligations by crops and the land's productivity for succeeding human generations. Bioregionalism made the land part of the community and its health a—or the—measure of right action. Rather than escaping from consumer society bioregionalists saw their action as a way to move the culture toward a relationship with the land we could, now, hardly imagine.

Bioregionalists found different ways to live with the land. Wendell Berry built a life around a nineteenth-century producer ideal and a rural community tied to farming, and he farmed with horses, but Wes Jackson, as committed as Berry to farming and a revitalized rural society, applied the knowledge of modern agriculture and land management to developing an ecologically responsible agriculture for the Great Plains. Trained in agricultural science, he established the Land Institute in one of Kansas's dying small towns and began experimenting with perennial crops, reasoning that plants that lived several years would obviate the annual plowing that contributed to soil erosion and more closely mimic the continuing productivity and soil conservation of natural grasslands. Bioregionalism, though, did not depend on agriculture. Gary Snyder did not farm, nor did Stephanie Mills, a California bioregional activist who followed a partner to Michigan and remained after the marriage foundered to work out a life in the northern forest as a bioregional activist and writer.[26] There were many niches around the edges of the industrial economy.

The central problem was not how to make a living but how to live, and, lacking established bioregional principles, bioregionalists used existing traditions as they searched for a new path. The tradition, as the cases of Gary Snyder and Wendell Berry suggest, need only be consistent and deep, for while they both explored the possibilities and limits of their choices before bioregionalism became popular, criticized the assumptions, values, and life of industrial America, and became bioregional examples for it, they built their lives on very different foundations. Snyder, a West Coast poet, spent several years in Japan (and much of that in Buddhist monasteries) before returning to the United States to make his home in the western foothills of the Sierra Nevada. Berry, on the other hand, wandered into modern academe but came back to rural Kentucky and to the evangelical Protestantism of his youth.

What they shared was a belief in the task of growing as humans by living in contact with the world, following its rhythms and respecting its processes, what Snyder called the "real work." They regarded living with and on the land as essential because the land provided the inescapably local context of life. "It is not enough just to 'love nature' or to want to 'be in harmony with Gaia.' Our relation to the natural world takes place in a *place*, and it must be grounded in information and experience." Living in that way "teaches you that you are in an interdependent condition with other beings, and it teaches you the sanctity of life, also how to take life; it solves, not exactly solves, but makes meaningful and beautiful the primary paradoxes that humans have to live with."[27] It allows you to see the sacredness of ordinary life by participating in it. "A subsistence economy is a sacramental economy because it has faced up to one of the critical problems of life and death: the taking of life for food. . . . Eating is a sacrament."[28]

Such an economy did not require doing away with electricity, running water, and mechanical power. Nor, despite Snyder's admiration for Native American traditions, did it mean living as the Native Americans had. It meant living the authentic human life, the one for which our biology intended us, a life in dialogue with the world around us. Doing that we would "make the world as real as it is, and . . . find ourselves as real as we are within it."[29] The "most beautiful of possible human experiences," according to Snyder, was to explore consciousness itself and become aware of the principles at work in the mind and the world. Exploration did not mean going to the mountaintops or living as a hermit. "The truly experienced person, the refined person, *delights in the ordinary.* Such a person will find the tedious work around the house or office as full of challenge and play as any metaphor of mountaineering."[30]

That life taught us to live on the continent, which called for even greater changes. For "the non-Native Americans to become at home on this continent, he or she must be *born again* in this hemisphere, on this continent, properly called Turtle Island." Making the land home required consciously accepting it as our home and our descendants' home "for millennia to come," which, he warned, would not be easy. We had no history of living with, rather than conquering, the land, and we could not rely on indigenous peoples. We had to find our place in the world. It would take time, but Snyder believed there was "no rush about calling things sacred. I think we should be patient, and give the land a lot of time to tell us or the people of the future."[31]

Like Snyder, Berry insisted on making a life within the possibilities and limits of one place, but where Snyder sought wisdom outside Western culture, Berry looked within. He believed that, as a society, we had to look within. Individuals could pick and choose from the great traditions of the world, but the majority of Americans were not going to become Buddhists. Any realistic plan for an environmental society had to build on familiar values. Since many environmentalists held Western ideas—particularly Christianity—responsible for our environmental problems, Berry had an uphill pull, but he believed that Christianity, including his own evangelical Protestantism, could guide us. Rightly understood, he said, Christianity did not devalue nature or justify our treatment of it. Our "destruction of nature is not just bad stewardship, or stupid economics, or a betrayal of family responsibility; it is the most horrid blasphemy. . . . The Bible leaves no doubt at all about the sanctity of the act of world making, of the world that was made, or of creaturely and bodily life in this world. We were holy creatures living among other holy creatures in a world that is holy."[32] To Berry, our life was not an exile or preparation for Heaven. We had a place here and the obligation to live with and nurture the world in which we live. Our work made up our prayer and connects us "both to Creation and to eternity."[33]

Our labor, though, only connects us to the world when social and economic relationships fostered full and fully human lives, and our current ones do not. The specialization of function in modern society produced a "calamitous disintegration and scattering-out of the various functions of character: workmanship, care, conscience, and responsibility" and reduced mutuality and cooperation to mutual interest.[34] In the last generation, the free market, "in which the freest were the richest," had displaced millions and killed communities. It was the antithesis of a healthy culture, which was "a communal order of memory, insight, value, work, conviviality, reverence, aspiration."

A healthy culture "reveals the human necessities and the human limits. It clarifies our inescapable bonds to the earth and to each other. It assures that the necessary restraints are observed, that the necessary work is done, and that it is done well. . . . For cultural patterns of responsible cooperation we have substituted this moral ignorance, which is the etiquette of agricultural progress."[35] To combat this, Berry turned, or turned back, to American agrarianism and small-scale community. He looked for a nation of neighborhoods small enough to allow individual contact and knowledge and in which land and wealth were distributed widely and evenly enough that individuals were largely self-sufficient within the larger frame of cooperative and community effort.

Snyder and Berry were not unusual in working toward radical change from an established tradition, only in using an explicitly religious (though in Snyder's case not theistic) base. Bioregionalists more commonly drew on ideas about spirituality in secular society of the kind Aldous Huxley developed in *The Perennial Philosophy:* a religion without a creed and mysticism without transcendence. Huxley looked below the level of creeds and declarations of faith in the world's religious traditions for common ground. He believed all these faiths rejected the modern belief in "the religion of Inevitable Progress—which is, in the last analysis, the hope and faith (in the teeth of all human experience) that one can get something for nothing." The search for meaning and purpose, he felt, had to concentrate on daily life and the ordinary world. The saint, he said, knew that all was part of God and did not reject the world for heaven, and "the doctrine that God is in the world has an important practical corollary— the sacredness of nature, and the sinfulness and folly of man's overweening efforts to be her master rather than her intelligently docile collaborator."[36] Huxley called for living in small communities, rejecting large-scale organization and large-scale loyalties, looking instead for decentralization and diffusion of power, encouraging widespread private ownership of land and the means of production, and seeking a renewal of local places and values.

The early activists, whether campaigning for green consumerism or making a living on the land, began, but only began, the search for a new relationship between humans and the land. They broke new ground, using conventional tools, and (more than people wanted to recognize) stayed within conventional limits. Green consumerism made people more aware of the consequences of their actions on the world, but worked within a system that treated people as consumers rather than citizens. To the tradition Huxley spoke of, bioregionalism added ecology, but left open the question of how six billion people were to live. Both approaches to the environmental life worked within a framework of ideas whose fundamental assumptions environmentalism rejected. People declared we must live as plain citizens of a biological community, but they used the language of individualism, autonomy, and choice, while ecology saw our ties to nature in terms of necessity. People looked for self-realization while ecology spoke of limits. It remained to discover what we needed to live as persons-in-community, constituted as much by relationships as independence and with as many responsibilities as rights.

MAKING SENSE OF PRACTICE

Environmentalists developed principles by acting on them, while trying to find the principles behind effective action. Academic philosophy gave little aid in the early years, for few philosophers found the movement interesting, and the most influential idea in environmental philosophy, deep ecology, worked against the academic grain that emphasized analysis and the rational ordering of principles. Established lines of radical thought had more impact, as environmentalists built on feminism and "left" political movements—anarchism, socialism, and communism—or debated the relation between these and nature's systems.[37]

These discussions, though often couched in political terms and only rarely in moral or spiritual ones, involved religious questions. As Jeffrey Ellis pointed out, arguments that the population explosion, or technology, or the hierarchical order of society were the "root cause" of environmental degradation meant we needed more than technical or social solutions: "Halting population growth, democratizing the technological decision-making process, restructuring society along nonhierarchical lines, and altering people's basic world views are not, by any means, simple solutions to the many deeply complex ecological problems that confront us. Each of these agendas taken alone would require nothing short of revolutionary changes in the ways Americans think, act, and relate to one another."[38] Each reached "people's basic world views"—who they believed people were and what place people had in the world. To answer those questions, environmentalists had to work out their views about knowledge and faith and had to fuse emotion and reason into a coherent worldview.

Deep ecology, which became an important popular environmental philosophy, grew because it fit the needs of the time. In a confused and confusing time, it emphasized the exploration of environmental ideas rather than a framework of logical propositions. Its founder, the Norwegian philosopher Arne Naess, saw deep ecology as a guide for people intent on finding their place in the universe, and his American disciples emphasized that perspective. In his introduction to Naess's *Ecology, Community, and Lifestyle,* David Rothenberg called deep ecology a reaction to a crisis that we feel by "immediate, spontaneous experience," a reaction that used ecological concepts to show us our true place as a species in nature and helped each of us realize our individual self, which was connected to all. He deliberately avoided defining the key term "self-realization" and presented Naess's ideas as a sketch for readers to use in coming to their own understanding.[39]

In 1985 Bill Devall and George Sessions wrote *Deep Ecology* for "the reader seeking a more authentic existence and integrity of character," presenting the new philosophy as "a theory of direct action which can help develop maturity," that the reader should make his or her own. Deep ecology was "a way of developing a new balance and harmony between individuals, communities, and all of Nature," a way that involves "working on ourselves, what poet-philosopher Gary Snyder calls 'the real work,' the work of really looking at ourselves, of becoming more real," and what Theodore Roszak describes as "the essential business of life, which is to work out one's salvation with diligence." They stressed the need for individual practice, the application of principles, for they regarded deep ecology as active, not a statement of principles but a guide to action. They approvingly cited Aldous Huxley's statement, from *The Perennial Philosophy,* that the "end of human life . . . is the direct and intuitive awareness of God; that action is the means to that end."[40] We would find that awareness outside the framework of the current society by returning to a self-regulating community and a "voluntary simplicity" which rejects the artificial complexity and unreal diversions of our modern lives.[41]

Naess and his followers called their practice "deep ecology" to distinguish it from the "shallow ecology" that looked to technical solutions and political changes, but they did not believe it reached the deepest level. In *Simple in Means, Rich in Ends*, Devall said environmentalists acted "in defense of the cosmos" and explored "the ideal state of being," and he characterized deep ecology's approach as "a profoundly objective spiritual way." Still, he quickly added that deep ecology was "not a new religion or cult, nor does it fight any religions." He described religions as "basic beliefs, religious and philosophical, from which a person derives his or her point of view in deep ecology" and deep ecology as a belief at a less basic level. People could, he thought, reach a commitment to deep ecology from any religious tradition. Some, he pointed out, held that "working from the teachings of St. Francis one can reach a deep ecology position from the Christian tradition." The goal, not the ground, took precedence. Regardless of their philosophical beliefs, people could unite around saving the planet.[42]

In deep ecology, environmental thought, like early environmental action, ventured into new territory along established trails. Deep ecology emphasized an individual approach to nature and an individual search for insight and denied any commitment to fundamental things. That made it appear as the kind of search rational people embarked on. But prudence did not bring people to environmentalism or deep ecology. Horror at the destruction of a world they felt was essential to our physical survival and our spirit did. Talk of self-realization and principles ignored people's primary concerns and deep feelings. If conventional faiths worked this way, they would seek converts by passing out theological works—and when was the last time someone came to your door with a copy of Calvin's *Institutes of the Christian Religion* or St. Thomas's *Summa Theologica?* Conventional faiths knew that commitment came before knowledge or understanding, and environmentalists, deep down, knew it too. There were, as Aldo Leopold said, "some who can live without wild things and some who cannot."[43] People believed that "in wildness is the preservation of the world" without philosophical reasoning or ecosystem theory. They felt it before or in spite of rational beliefs.

FINDING A MYTH

Environmentalists' attempts to live a new life, either inside or outside American society, exposed the contradictions in their developing position. They rejected some central American values from a stand strongly affected by those values. Environmentalists needed the authority, facts, and perspective science provided, and used them, while fighting against the emotional detachment and political neutrality scientific work seemed to demand. In every environmental issue, from Carson's case against DDT to global warming, environmentalism relied on scientific information, and at a more basic level on ecology's description of the world as interconnected, dynamic systems.

Ecology defined environmentalism's idea of a community and measured its integrity and stability—if not its beauty. Ecology's descriptions of relationships allowed environmentalists to connect human action in California or Washington to changes on the Great Plains or in the Amazon rainforest. Its theories showed the apparently separate problems of air and water pollution, dwindling wilderness, and vanishing wildlife as parts of the same problem. Training and practice, though, taught scientists to stand apart from what they studied, while environmentalism demanded personal, emotional commitments to the processes of life, even identification with the natural world. Where science saw reproducible, "objective" data as the only real knowledge, environmentalism insisted that personal experience had value. Science, as science, looked for facts; environmentalism looked for meaning.

The ideal of objectivity, though, was crumbling even as environmentalism developed. German scientists' collaboration with the Nazis gave "detachment" a bad name, and the dropping of atomic bombs on Japan shook American physicists' confidence in "objectivity." Radioactive fallout in milk, the birth defects in several thousand babies caused by an apparently harmless drug (thalidomide), and other, less spectacular disasters raised concern about the consequences of science and scientists' stand both inside and outside the scientific community. Between the publication of *Silent Spring* and passage of the National Environmental Policy Act in 1969, attitudes clearly shifted. In 1962 scientists hesitated openly to support Carson. Robert Rudd's *Pesticides and the Living Landscape* (1964) scrupulously avoided emotional language, at least in part because one of the referees insisted the case be made with "just the facts."[44]

Charles Wurster, who led the Environmental Defense Fund's scientific team in the DDT case, came under fire from academic colleagues who felt scientists should not be "engaged," and the Environmental Defense Fund found it hard to get scientists to testify except in the neutral forum of legal proceedings, where they could present themselves as expert witnesses. By the late 1960s, some ecologists and biologists—most notably Paul Ehrlich, Barry Commoner, and Garrett Hardin—openly broke with the social norms of "objectivity" and "detachment" and joined the environmental cause. Twenty years after that, when concerned scientists formed a new field, conservation biology, to apply scientific research to ecological restoration and preservation, the old ideas were dead.[45]

Environmentalism also had to go beyond science's common perspective of a material universe of laws. That had a certain grandeur but more often brought a cold chill. Cosmology showed the Universe's beauties (photos from the Hubble Space Telescope appeared in coffee-table popular science books), and physicists and cosmologists spoke of reading the mind of God, but they brought no messages of hope and found no consciousness or intention. Darwinism showed a bleak world of struggle and death; while Robinson Jeffers found beauty in "the wolf's tooth [that] whittled so fine / the fleet limbs of the antelope," even he found little comfort in it. Nor did others. Few found exhilarating the prospect of making our own purpose in an indifferent universe.

In trying to show nature's grandeur, science chilled the blood, and the American civil faith, while it offered material rewards and inspiration, held out no higher purpose than being a good consumer, and dying in battle (nationalism's equivalent of martyrdom) had limited appeal even to the age group most likely to do it. Religious traditions persisted in part because they offered people the chance to participate in a drama larger than their lives or even humanity. Christianity, to take an example close at hand, put everyday lives and everyday actions in the context of the ultimate battle of Good against Evil. In language and metaphors that inspired several centuries of artists, Christianity described this struggle and its even more spectacular end, Judgment Day, when all would come before God, and the unrighteous, with the Devil and all his minions, would be cast into the lake of unending fire while the righteous were raised to the throne of God.

Not by coincidence, the secular belief most commonly described as a religion, Marxist communism, had its own great drama, the struggle between the social classes, in which all had a part. As in the Christian scheme, evil was the corruption of good. Marx held that capitalism had to be swept away but admitted it had played a necessary role and even praised its contributions. The wickedness of capitalism lay in failing to give way before its anointed successor, the working class, which embodied the new economic system—socialism. Like Christianity, Marxism included everything in salvation history. Reading the signs of the times in the light of Marx's writings, the good Marxist knew what each situation required, what was good, and what had to be consigned to the dustbin of history. The Marxist faith included political beliefs, literature, recreation, even—as George Orwell noted—table manners and pronunciation.[46] It offered an Earthly Paradise with the apotheosis of the workers and the withering away of the state. Operationally, as the philosophers would say, Marxism formed a religion, and for a century and a half millions around the world found it a fighting faith to which they committed their lives.

In the early years environmentalism built its own still-incomplete myth, an abbreviated salvation history in which the crisis of the late 1960s constituted the "moment of revelation" and Americans' response foreshadowed a turn to repentance and the beginning of the passage to heaven. That scientists rather than prophets brought the news that the human species threatened the very processes of life—and that the revelation came as scientific papers and not stone tablets or burning bushes—made no difference. The Word had come, and preservation of earth's ecological processes and the diversity of life became a sacred cause, saving the planet an ultimate Good, and the human destruction of nature's processes Evil. In wilderness the movement found a place for mystics in the movement and offered in recycling, energy conservation, and the like practices for the ordinary believer, and in bioregionalism possibilities for a new life. The possibility of a civilization where people could live fully human lives in a community serving and preserving earth's life served as heaven, the environmental collapse if we failed to act, hell.

The environmental myth addressed an issue at the center of most people's religious beliefs (and of concern even to those who did not speak of any) but one secular faiths avoided—life after death. "Religion," said William James, "in fact, for the great majority of our own race *means* immortality, and nothing else."[47] Environmentalism did not believe our personality or consciousness survived the body's death, but it did hold out the virtual immortality of our bodies feeding earth's cycles. That we returned to the earth everyone acknowledged, even if they denied ultimate significance to the fact. Christians looked to the Resurrection, but their burial service said, "Ashes to ashes, dust to dust."

Romanticism found in our decay a secular transformation that united us to life. Look for me, said Whitman, under your boot soles, and Robinson Jeffers and Edward Abbey spoke of our passing on into the lives of hawks or vultures. Twentieth-century Americans cremated bodies and returned the remains to places with significance for the deceased: for sports enthusiasts golf courses and football stadiums, for Civil War buffs battlefields, and for those committed to the conquest of the universe the rocket ride into outer space or to the moon. Environmentalism gave biological significance to the act of returning our phosphorous to the planetary pool. Although the final destination was the same, no one asked to be buried at the local sewage farm or expected to have their ashes flushed. Instead, ashes were with some small ceremony distributed on family land (though rarely a suburban front lawn), in forests, on mountains, or in national parks. The sea, a symbol of the cradle of life and many other things, was popular even among those who did not embrace environmentalism.

The first generation of environmentalists, not surprisingly, only began to build the movement, but it began, despite the culture's silence on ultimate questions, opponents' use of the term "religion" to discredit it, and its appropriation by every "movement" of the 1960s. Environmentalism rejected the term "religion," but environmentalists, often unconsciously, laid the foundations of a myth of their worldview—a description that showed people the world, their place in it, and their duties—and pointed to a spiritual search in nature. It is easy to make fun of early environmental discussions and practice. Debates over saving energy and resources had an intensity and earnestness all out of proportion to results, and no one with a sense of humor or perspective could take seriously pronouncements about the imminent transformation of human consciousness.[48]

Environmental passions, though, were neither simple nor simple-minded. People recycled materials, saved energy, and bought green products because they wanted to save the earth's biological systems from destruction by industrial society. The movement changed—even if it did not transform—American culture. The Reagan administration, confidently expecting to sweep away environmental laws, agencies, and programs as remnants of the sixties, found a large and surprisingly bipartisan opposition to its plans, even in areas where the economies depended on primary production. Environmentalism formed plans for a responsible society and an environmental way of life that, even if it could not be realized immediately (what large, good dreams could?), at least faced up to real problems the conventional wisdom dismissed and deep human needs the society failed to acknowledge. Warnings of environmental disaster were not pleasant, but they were more realistic than ideas of endless technological progress.

In the 1980s environmentalism reached a crossroads. Despite all it did, pollution continued, population grew (though a bit more slowly), extinction rates rose, and wilderness dwindled. Some environmentalists said the movement had failed, and a strong, deep, and sometimes bitter debate developed within the movement. Beneath these visible changes ran less visible, more fundamental ones. In the early years, when it seemed that industrial society must soon collapse or be transformed, environmentalists said little about the long term, but as the years passed without the coming of the revolution or the apocalypse environmentalists, finding themselves in the position of early Christians wondering why the Second Coming was on hold, began more consciously to look beyond the inspiration of the moment for a way of life that would guide and inspire them through a lifetime. That search involved deeper questions—about what "individual" and "community" meant in a world that ecology described and how a society that respected nature's limits would define these things. If we could not be autonomous individuals shaping nature, what could we be? The next chapter takes up environmentalism's struggle to develop its vision, a struggle still going on, one in which the movement's relation to religious questions is central.

5 / In for the Long Haul: Living in the World

Environmentalism began with crisis and a sense of commitment. People rallied against DDT, oil spills, dying lakes, and vanishing wildlife; they recycled cans and newspapers; they stuffed insulations in every crack; they joined organizations; and they published newsletters. Despite their efforts, aided by bell-bottomed jeans, sex, drugs, rock and roll, the movement of the sun into Aquarius, and the ritual burial of a few automobiles, human consciousness did not change, social values remained much the same, the industrial machine ground on, population grew, and wilderness shrank. Local disasters came with unhappy regularity, but the world's ecosystems did not visibly collapse.

The foot soldiers of environmental protest—the college students who came to the streets for peace and the environment and against the Vietnam War—grew older. They changed their clothes and cut their hair, swapped Rex's bandana for a dog collar and took him to the vet, got jobs, got married, bought a house in the suburbs, and had two (the then environmentally correct number) of children. Fervor declined; commitment, however, remained. Environmentalism found a place in the political dialogue, and environmental organizations moved from the streets to the legislature and on to the anterooms of administrators, where they watched over appointments of assistant undersecretaries and reported to their members on the iniquities of the current administration. Environmental ideas spread and took root, as the Reagan administration discovered when it tried to roll back environmental protection; soon even anti-environmental politicians declared their commitment to "sound" environmental policies.

By the mid-1980s, the movement seemed at an impasse. It had changed the society in some ways and secured a place in political and social debates on every level, from town councils to presidential campaigns, but even James Watt's open and blunt opposition during his term as secretary of the interior in the early 1980s produced only a new wave of members for the Sierra Club. Environmentalism, which in the 1970s seemed on the verge of transforming the society, seemed in the 1980s only another interest group. Since, by the logic of its position, the movement had to change social values or watch nature die, many charged it had failed. Radicals accused established organizations of "selling out" and turned from politics to direct action. Others believed the movement had lost touch with its roots and failed to grasp the central cause of environmental degradation, which they variously identified as overpopulation, technology, the wrong use of technology, science, the wrong kind of science, or the social and economic structures that perpetuated (choose one or more of the following) hierarchy, racism, patriarchy, poverty, and the postcolonial domination of the Third World by the First.[1]

Each faction saw a clear road ahead, but the movement as a whole did not, for the environmental crisis—a situation that would soon resolve itself or be resolved—had become a slow-motion disaster, something we had to live with as we changed it. That had consequences for environmental politics, but also for basic beliefs, forcing a different approach to questions of finding, exploring, and living out environmental principles. It was one thing to live your principles secure in the knowledge that they would soon triumph or society would soon collapse, another to do it with no assurance of victory and no hope of a quick resolution. In addition, even if industrial society collapsed in the future (with consequences as deadly as those predicted by the worst doomsayers), in the short run life was not particularly uncomfortable. To face this environmentalists needed not just principles but ways of living them over a lifetime, ways that would help them form emotionally significant and satisfying connections to nature while living in a hostile or indifferent society.

Environmentalism's inheritance and even the first generation of activism offered little guidance here. Conservation and nature preservation accepted society. They used the political and administrative machinery of government—the one to make efficient use of resources, the other to protect areas of scenic beauty for recreation and refreshment—and saw nature as only a part of a life well lived. In exploring a new life in nature environmentalists had begun, and often ended, within the familiar framework of individual choice and individual action. Green consumerism looked for individuals to build a sustainable society by enlightened decisions and regarded citizenship in the biological community in terms of choice rather than necessity. Bioregionalism began with people choosing places and ecosystems, and rural and often agricultural lives left open the question of how all of us were to live and on what principles we were to build our lives.

Neither path faced squarely the problems of living and forming new values over a lifetime, the problems involved in enlarging the community to include all forms of life and recognizing the extent and nature of our ties to local and global ecosystems. People thought of Leopold's land ethic as an answer. Under our circumstances it presented questions about what it meant to be an individual in this kind of community. American society saw the world as a collection of separate things autonomous individuals used as they wished, community as something people formed by giving up some freedom for the benefits of association, and they identified freedom with the ability to use or use up nature. What did "individual," "community," and "freedom" mean in an interdependent world? If the health of the community came first and if individuals existed, as they seemed to, by virtue of their function in the community, what did autonomy and freedom mean?

Answering those questions required that we examine our understanding of how we should live and what we should live for and how societal and economic structures served that end. Environmentalism faced these from its start, but crisis, hope, and fear obscured them in the early years. E. F. Schumacher's *Small Is Beautiful* focused on what makes for a good human life. The need to change in fundamental ways lay at the root of the Club of Rome's study, *The Limits to Growth*. Working out an environmentally responsible life became more urgent by the mid-1980s, but environmentalists found little help in theory, for their questions had less to do with principles than the application of principles. In these efforts environmentalists benefited from the American emphasis on practical action (or, less kindly, anti-intellectualism), for they were engaged in what was formally called "praxis": the practice of virtue combined with reflection on it. "Truth," in that formulation, was "not only to be thought but to be lived and done, and the living and doing of truth is a condition of grasping it."[2] William James described that attitude in *The Will to Believe*. Some propositions, he said, would not reveal what they were unless we assented to them, took them as true, and acted on them. Only in living the environmental life could people find out how to live it and discover what satisfactions it would yield.

Living the environmental life began, but did not end, with confronting growth in a limited world—began because growth seemed an immediate threat to the world's ecosystems, could not end there because analysis led from the foundations of economics into the values behind it. Beyond that debate we have to look at environmental action, which took one of two paths. One was the search for nature around home and in daily life, which often became the individual quest for wildness rather than wilderness, the other the slower and more social task of finding scriptures that expressed environmentalism's deepest convictions and symbols that bridged the gap between people and the lands on which they lived. The first challenged the boundaries people normally drew between nature and culture. For a century, Americans looking for the good of nature and in nature for insight into reality moved further from society. Thoreau found wildness in Concord's woods. Muir fled to the fastnesses of the Sierra Nevada. The wilderness movement emphasized areas apart from industrial society. Now nature's defenders asked how we might find nature and wildness in the spectrum of lands that ran from the backyard to the Bob Marshall Wilderness, even as they lived in a society that surrounded and threatened wilderness. That search probed and confused conventional views of nature and culture and those inherited from nature appreciation as well.

The second approach, environmental virtue, developed within the movement, for nature appreciation had not discussed virtue. Environmentalists began by preaching the need for humans to be humble before nature, but as the first wave of enthusiasm waned in the 1980s they asked more clearly what other qualities their cause required, raising direct moral and religious questions about what made for human goodness and how people should live their lives. Searching for nature and for appropriate human qualities both involved the developing environmental critique of Americanism, the country's civic religion.

STRUGGLING WITH GROWTH

Environmentalism challenged the American faith that continuous economic growth fueled by individual action brought social prosperity. The leather-bound, gilt-edged editions of Adam Smith's *Wealth of Nations* that high-level Reagan administration appointees reportedly kept on their office shelves testified to faith in a creed about the world and the righteous as clearly as the white-leather-bound, red-lettered editions of *The Living Bible* on the end table next to the plastic-covered sofa. Economic growth, founded on free markets and free individuals, would turn more and more of nature's endless bounty into things, enrich us all, guarantee liberty, and make us happy. The process had no end, for human ingenuity would overcome any obstacle of materials or energy in nature, providing us all we could ever need or want. Environmentalism saw these beliefs as destructive and self-destructive illusions. Continuing expansion required that the world have no limits, and ours had them. Conventional values measured everything in dollars; environmentalists believed nature had value in itself and was essential to our lives and spirits.

This conflict of values appeared in the first public arguments for environmental policies in the 1960s. Garrett Hardin addressed growth and markets in his article "The Tragedy of the Commons" (1968), but it was the Club of Rome's computer model of the world system, *The Limits to Growth*, that made limits a major topic of public debate by putting the prestige of numbers and statistics and the magic of the computer behind its description of a limited world.[3] Researchers created a computer model of the world system by identifying the "five major trends of global concern—accelerating industrialization, rapid population growth, widespread malnutrition, depletion of nonrenewable resources, and a deteriorating environment"—and using the statistical record since 1900 to write equations linking the variables and charting their changes.[4]

The equations yielded curves representing change from 1900 to 1970, which the computer extrapolated into the future. That produced a grim picture, for the equations showed that if we continued on our present path, the human race would reach the world's limits within two centuries. Then people would begin to starve or die of disease or pollution. The researchers changed the variables. They assumed more resources, less pollution from each unit of production, more efficient farming, soil conservation, and better methods of birth control, but none of these factors—or any combination of them—changed the outcome. Some calculations staved off disaster, but only for a generation or two, then another part of the world system would collapse. We could not, they concluded, blame technology or pollution. Our problems stemmed from pursuing growth in a limited world.

Because this was the first computer model of the earth's systems, no one could say how accurate it was, but every news organization could say how good the story was: the computer—in those innocent days the very voice of truth—said we were doomed. An academic study became front-page news and the center of a storm. Donella Meadows, one of the authors of *The Limits to Growth,* recalled that criticism came "from the left and the right and the middle. The book was banned in the Soviet Union and investigated by President Nixon's staff. The Mobil Corporation ran ads saying 'growth is not a four-letter word.' Disciples of Lyndon LaRouche and the National Labor Caucus picketed our public appearances. Mainstream economists competed with one another to see who could write the most scathing reviews" (these appeared in a volume entitled *Models of Doom*).[5] What Meadows found particularly disheartening was the emphasis on disaster. *The Limits to Growth,* she later protested, "was about choice." The group did not write to "predict doom but to challenge the myth of growth as the answer to all problems."[6]

In challenging the "myth of growth," though, *The Limits to Growth* struck at the heart of the modern faith: the belief that human reason could understand and so master the world and free us from nature's limits. On

this point capitalists and communists joined hands. American social policy depended on economic growth to solve social problems. We did not
need to rob the rich to help the poor; the expanding economy would help
all—or at least anyone willing to work. Communist regimes, though they
rejected the magic of the marketplace, believed as devoutly in the cor-
nucopia of the factory. Scientific socialism was going to build the Earthly
Paradise without the pain and injustice capitalist economic development
inflicted on the workers. Everyone condemned *The Limits to Growth* for
the same reason they hurled anathemas at Thomas Malthus's *Essay Concerning Population:* it questioned core doctrine—the godlike power of
human reason and the godlike human destiny to conquer nature.

The Limits to Growth set off a debate that could not be resolved, for
the two sides began from different descriptions of the world and humanity's place in it. Environmentalists saw humans as limited creatures in a
limited world who, in the long run, had to live with and within the world's
biological realities. Their opponents believed human intelligence allowed
us to overcome nature's limits; humans were, as the title of one of Julian
Simon's books had it, "the ultimate resource."[7] Because discussion of
growth took place in public, the only things that counted were things that
could be counted: market prices, average life expectancies, projected food
supplies, arable land, reserves of energy, and the monetary value of
nature reserves.

Environmentalism as a reform worked at this level, but environmentalism as an examination of humans' place and purpose in the world could
not accept economics as a measure of life. When environmentalists
accepted economics they argued from a position that implicitly dismissed
their concerns; when they ignored economics, society ignored them. Arguing on their opponents' ground troubled them, but they saw no way around
it. Daniel Botkin asked why "people who love nature tend to resort to material justifications for conserving it" when arguing from economics constituted "an implicit retreat from a belief in the power of ideas and the
power of human passions" and reduced environmentalism to social and
political action, rather than something that struck "at the very heart of
our existence and the heart of the human condition, affecting every part
of our lives."[8] He answered his own question by saying that environ-

mentalists felt these were the only arguments people would listen to—which neatly pointed to their dilemma.

Some tackled it by introducing environmental realities into economics. They replaced the idealized world of the market (exchanges between willing buyers and sellers with perfect information in a situation where no one could set prices) with a model incorporating the "externalities" economists ignore: the effects of market action that no one paid for, like polluted streams, dead fish, and stinking air. They attacked that other simplification of classical economics—the world as infinite source and infinite sink—arguing that this view, plausible in Adam Smith's day, was now obsolete. We could not, they said, ignore the dead rivers, depleted fisheries, and ruined land that came from mining the soil and dumping the costs of industrial production into the air and water.[9]

Putting nature into economic analysis, though, only went halfway, for economics had foundations environmentalism could not accept: human beings as fully autonomous agents, all values reduced to the single standard of money, and human welfare measured solely in economic terms. Environmentalists knew about these problems, for Schumacher analyzed them in *Small Is Beautiful.*[10] In the early years, Schumacher's admirers often emphasized peripheral arguments like the decentralization of industry and appropriate technology for the Third World rather than the book's critique of classical economics' philosophical foundations. These are worth reviewing, for they showed the faith on which the conventional view rested.

Schumacher began by dismissing the idea that the laws of economics were laws of nature. They were, he said, human constructions based on a human understanding of the world. Worse, they did not reflect reality, for they assumed that society could have peace and happiness if individuals simply worked to increase their possessions. "An attitude to life which seeks fulfillment in the single-minded pursuit of wealth—in short, materialism—does not fit into this world, because it contains within itself no limiting principle, while the environment in which it is placed is strictly limited." That attitude was central to the field. It was "inherent in the methodology of economics to *ignore man's dependence on the natural world.*" That view produced "a system of production that ravishes nature and a type of society that mutilates man." Ours is a "philosophy of materialism, and it is the philosophy—or metaphysics—which is now being challenged by events."[11]

Schumacher contrasted conventional economics with a system in which the "meta-economic basis of western materialism was abandoned and the teaching of Buddhism put in its place." (He did not, he said, have to use Buddhism; the "teachings of Christianity, Islam, or Judaism could have been used just as well as those of any of the great Eastern traditions.") The Buddhist saw "the essence of civilisation not in a multiplication of wants but in the purification of character" and measured development, then, not by goods but by liberation. That did not mean rejecting technology or taking spiritual goods rather than economic ones, but it did mean economic efficiency could not be the sole measure of success, and people came before production. Development had to be assessed by the way it nourished humans, and the land seen as something more than a place for production. From that point of view, the fundamental idea of property took on new form. Ownership, Schumacher believed, could not serve individuals alone; it had to serve society as a whole—a position the conventional wisdom saw as blasphemy.[12]

This radical case rested on a deeply conservative strain of the Western tradition, which retained the culture's important values in the face of change. Schumacher relied on a position laid out by Edmund Burke, who in the 1790s argued that society was an organic growth whose laws and traditions we should respect and change only slowly and with care. He also drew on a line of agrarian criticism from the same era that blamed markets and the factory system for destroying established communities and values. He did not want primitivism or an agrarian society, but he did seek a return to what we knew.

Schumacher couched his appeal in religious language and ideas. "This clear-eyed objectivity [requires some] silent contemplation of reality during which the egocentric interests of man are at least temporarily silenced. The guidance we need for this work cannot be found in science or technology, the value of which utterly depends on the ends they serve; but it can still be found in the traditional wisdom of mankind. . . . It is hardly likely that twentieth-century man is called upon to discover truth that had never been discovered before." Quoting Christ's call to seek first the kingdom of God, Schumacher continued: "In the Christian tradition as in all genuine traditions of mankind, the truth has been stated in religious terms, a language which has become well-nigh incomprehensible to the majority of modern men." Like Huxley in *The Perennial Philosophy*, Schumacher pointed to a core of common philosophical beliefs and a common stock of virtues. Having used Buddhism as a model, he recommended for environmental virtues those Christianity had borrowed from stoicism and made the cardinal virtues: prudence, justice, temperance, and fortitude (or courage). Among them he stressed prudence, which he saw as the understanding that to find the good, we had to begin by knowing what was real.[13]

WILDNESS RATHER THAN WILDERNESS

As environmentalists became more concerned with the long term, some took up Schumacher's intellectual attack, but more commonly they worked out ways of living by practice and reflection. Exploration served better than thought because there was no obvious way from the general proposition that we should be humble before nature to the actions of daily life that healed the land. People could (and in the early years many did) reverse the idea of conquering nature and demand that we simply leave all to nature, often repeating Barry Commoner's laws of ecology, particularly the dictum that "nature knows best." A decade later they embraced James Lovelock's Gaia hypothesis in which the earth's chemical and biological processes formed a stable, self-correcting system favoring life, and so all we need do is rely on the earth.[14] That stand, like conquest, proved too simple for real life. The question remained; what were we to do and how should we guide our lives?

Society suggested many things people could do, but it had little to say about why they should do them. Schumacher lamented that the terms and ideas of religion were "incomprehensible to the majority of modern men," but so were common views of political and social action. Environmentalism could not present its arguments in a public arena that served (as it had to in modern America) as a neutral forum where groups contended for political power without abandoning its belief that nature had value. It linked individual conduct and character on the one hand and what society needed on the other, whereas the American political model saw factions contending selfishly for their own interests. People could, and did, seek to bring nature into society with the language of rights and the forms of legal representation, but these, as much as economic arguments, abandoned the movement's core beliefs. Environmentalism's demand that people act as citizens, taking responsibility for the functioning of the community, and not act as consumers, making choices in a market, also made communication with many people harder.

As the environmental crisis turned into the environmental situation, environmentalists began to consider what it meant to be an individual-in-community and what virtues and values that life required.[15] Without forgetting wilderness or the ongoing destruction of the earth, they explored the lessons and possibilities in the mixture of wild and tame, nature and culture, where we live our lives. That program ran counter to a century-long movement toward searching for insight in ever more wild areas. Thoreau found the source of life in Concord's farms, woodlots, hay meadows, and abandoned apple orchards. Fifty years later, Muir looked to the high mountains, railed at commerce's intrusion into the holy domain of the Sierras, and considered people to be blots on the landscape. Another half century, and environmentalists enshrined wilderness as nature pure, a place of pilgrimage where we could avoid society and its works. In looking for insight and lessons in tamer areas, environmentalists blurred the sharp distinction they, and anti-environmentalists, made between nature and culture. The explorers of the semi-wild did not reject wilderness, want to do away with wilderness areas, or even see wilderness as less important—though wilderness's more ardent defenders made all those charges against them. They did see wilderness as one end of a spectrum and against the belief that nature was something we visited, leaving only footprints, taking only memories and pictures; they looked for the wild everywhere.

Michael Pollan's *Second Nature: A Gardener's Education* (1991) showed that break with the wilderness tradition and the ideas behind it. Our "habit of bluntly opposing nature and culture," he said, ". . . has only gotten us into trouble and we won't work ourselves free of this trouble until we have developed a more complicated and supple sense of how we fit into nature." Wilderness alone would not serve, for there were "many important things about our relationship with nature that *cannot* be learned in the wild." From wilderness we learned "more than we needed to know about virginity and rape, and almost nothing about marriage." Wilderness as a concept often hindered thought more than it helped. Wilderness appeared in environmental writings as the market did in anti-environmentalist tracts. To its devotees, each concept seemed a "quasi-divine force . . . that, left to its own devices, somehow knows what's best for a place."

Pollan wondered "if perhaps the wilderness ethic, for all that it has accomplished in this country over the past century, had now become part of the problem. I also began to wonder if it might be possible to formulate a different ethic to guide us in our dealings with nature, at least in some places some of the time, an ethic that would be based not on the idea of wilderness but on the idea of a garden." As a symbol, the garden was one of the oldest and richest in Western culture, and, as physical space, it offered active engagement with nature through our bodies, the kind of connection industrial society had broken.

Working in his garden and reflecting on the relation between the ideal in his mind and the messy reality on the ground, Pollan came to see gardening as a way to integrate nature and culture—a fresh approach, accessible to all, between the extremes of pristine wilderness and conquered nature. The garden already was for "most of us our most direct and intimate experience of nature—of its satisfactions, fragility, and power," and here, where people cooperated with nature's processes, we might look for "forms of human creation that satisfy culture without offending nature. . . . The idea of a garden—as a place, both real and metaphorical, where nature and culture can be wedded in a way that can benefit both—may be as useful to us today as the ideas of wilderness have been in the past."[16]

Some people tried more directly to fuse environmentalism with daily nature by—as the subtitle of Sara Stein's *Noah's Garden* put it—"restoring the ecology of our own back yards."[17] Both gardens and ecological lawns drew on the agrarian tradition of good stewardship, which saw our rela-

tion to the land in moral terms, but instead of a landscape of production—
the earth yielding its fruits to the careful husbandman—they imagined a
landscape of diversity, where the earth yielded its fruits for all forms of
life, and these approaches included themes of atonement and reparation,
which farming did not. Both brought nature and culture together to make
a world where humans, human-shaped landscapes, wildlife, and wild
ecosystems existed side by side.

The restoration movement consciously rejected the extremes of suburban life, when "development" meant cutting down all the trees and grad-
ing every inch of ground before building houses, and afterward putting
in lawns maintained by chemicals and trimmed by machines. It wanted
to bring back the vibrant, diverse world the bulldozers had scraped away.
Restorationists' enthusiasm made native plants, "natural" lawns, "bird-
friendly" backyards, and "butterfly sanctuaries" a strong niche market by
the 1990s. In Texas, for example, you could buy native buffalo grass sod
for the (still mandatory) suburban lawn, native shrubs and flowers for
landscaping, and native wildflower seeds for the garden in back (the companies, unfortunately, offered nothing to protect the natives from the
exotics that flourished in the irrigated suburban ecology). Companies from
Texas to the Pacific offered xeriscaping—landscaping with local plants
that did not need irrigation—which allowed people to show ecological
responsibility and enjoy lower water bills. Novelty and nature appreciation drove the boom, along with a genuine desire to participate in the
drama of remaking the world and atoning for our species' sins, if only by
changing a backyard. Restoration looked to a domestic Earthly Paradise,
where nature and culture met and mingled and humans, instead of impos-
ing their ideas on the land, learned what the land would support and
cooperated with it to realize its possibilities.

Amateur natural history served as another avenue into part-wild
nature, and books like Robert Pyle's *The Thunder Tree: Lessons from an
Urban Wildland* encouraged a renewal of the amateur tradition.[18] Pyle,
following a long tradition, saw nature as a doorway to great truths. When
teaching children about butterflies, he often placed the insects on their
noses, which delighted them. "Somewhere beyond delight lies enlightenment. I have been astonished at the small epiphanies I see in the eyes

of a child in truly close contact with nature, perhaps for the first time."[19] To lead children to those epiphanies and beyond, he urged local action to save small patches of ground where children could collect butterflies, catch frogs and fish, and generally mess around without adult rules or supervision. We needed these areas, he argued, if we wanted a culture that lived with the land, for they served as "places of initiation, where the borders between ourselves and other creatures break down."

It was through "close and intimate contact with a particular patch of ground that we learn to respond to the earth, to see that it really matters," and for most of us that was some unspectacular spot in our neighborhood. Pyle had found nature along the High Line Canal east of Denver, and he traced the history of the canal and its relation to Denver, its impact on the people who relied on it for irrigation water, its eventual abandonment, and his efforts to preserve it. "Had it not been for the High Line Canal, the vacant lots I knew, the scruffy park, I'm not certain I would have become a biologist. . . . The total immersion in nature that I found in my special spots baptized me in a faith that never wavered. . . . It was the place that made me."[20]

The "Nature" and "Environment" shelves in bookstores proved Pyle was not a voice crying in the social wilderness. Before World War II, these shelves had held only field guides to birds and, after World War II, very elementary guides to common forms of life and to the night sky. The explosion came in the 1960s. Guides appeared treating every kind of visible thing in nature—trees, flowers, butterflies, insects, mammals, even such despised creatures as snakes and amphibians—and then new kinds of guides that dealt with ecological spaces like seashores, tidal zones, Eastern forests, and deserts, or told readers how to make nature observations "everywhere— from an untamed wilderness to an urban landscape."[21] John W. Brainerd's *Nature Observer's Handbook* discussed ecological areas in a section titled "Nature's Patterns," and then, in "People's Patterns," directed readers to nature on bicycle paths and in quarries and mines. It recommended pastimes, such as urban rat-watching, that fell well outside conventional nature appreciation.[22]

Robert Sullivan's *Meadowlands: Wilderness Adventures at the Edge of a City* mixed nature and culture even more by applying the conventions and language of nature literature and the wilderness journey to the Meadowlands, a marshy area just across the Hudson from New York City, bordered by chemical plants and laced with freeways. Sullivan approached this "thirty-two-square-mile wilderness, part natural, part industrial," as an "undesignated national park." Having resisted development over centuries of white settlement, the area became, by negligence, exploitation, and its own persistence, land that, once again, we could explore. In it we could hear "a kind of wild industrial New Jersey sound track, which, unlike the environmental sounds of the Eastern forests and the Pacific Coast whale migration routes, is not available on cassette or CD" (certainly the Nature Store would not sell the cries of herons, the lapping of the tide, and the swishing of sawgrass mixed with the sounds of freeways and machinery).

Sullivan described his trip to an area called "Walden swamp" as a wild journey, but in terms that mixed nature and culture. He wrote of a "three-inch-thick rusted cable sticking up out of the water like a water snake" and an "egret, its long curved white neck the shape of a highway off-ramp, its white feathers the color of Styrofoam," flying off toward the smoke-stacks of Newark. Visiting the garbage piles, he found "a little leachate seep" a few yards from "a benzene-scented pool [where] a mallard swam alone." The seep would eventually mingle with the groundwater, but now it was "pure pollution, a pristine stew of oil and grease, of cyanide and arsenic, of cadmium, chromium, [and] copper."[23]

Sullivan unsettled ideas about nature and culture by looking for nature in the Meadowlands, describing pollution in language associated with nature, and comparing the natural and the man-made, but he retained the canons of nature writing that went back to natural theology. Like mainstream volumes, it recounted an individual search for insight carried out on a physical journey that led to meaning and beauty, and it used nature writing's established language and conventions. They were flexible. Muir had used them in describing his discovery of a beautiful flower in a Canadian swamp, Annie Dillard, the sight of a giant water bug sucking the insides from a frog.[24]

Nature writing's conventions did not require agreeing on the message, just on the possibility of finding meaning in nature. Stephen Jay Gould

and Stephen Weinberg could use them to speak of a world without purpose and place for humans, Ursula Goodenough and Chet Raymo, to
described a universe that was our home, because all accepted the premise that the facts of the universe led on to meaning.[25] The meaning might
be that there was no meaning, but even the non-message could be read.
Sullivan only departed from the main trail by looking for message and
meaning in what did not seem to be nature.

SCRIPTURES AND TOTEMS

By the 1980s environmentalists had incorporated the older nature literature as part of their own. No council defined the canon, but the shelves
labeled "Nature" or "Environment" in any large bookstore included the
great works and the great authors. The line began with Thoreau, whose
life had been the canonical example of the search for truth in nature, his
shack and Walden Pond a place of pilgrimage, and *Walden* a holy text,
long before environmentalism had begun. His message continued to attract
even hardheaded environmentalists. E. O. Wilson—no starry-eyed
idealist—began a book on preserving nature's diversity, *The Future of Life*
(2002), with a letter to Thoreau, telling of his visit to Walden Pond and
enlisting Thoreau in his cause.[26]

John Muir came next, a figure whose reputation, like Thoreau's, continued to grow. The Sierra Club, which he founded, made him a local
saint and the Sierra range an outdoor temple to his memory in the early
twentieth century, the patron of wilderness as that movement developed,
and the exemplar of resistance to industrial civilization for environmentalism. His acolytes battled the official clergy, the National Park Service,
over the care and preservation of the sanctuary—Yosemite Valley. The
Library of America enshrined his work, as it did Thoreau's, in its chaste
volumes, but the cheap paperbacks of *My First Summer in the Sierra* were
a better index of his popularity.[27] Like Thoreau's, Muir's work and life
generated a minor academic industry. Scholars parsed his works, debated
his values and how they developed, and discussed how and when—or
whether—he discarded the Christianity of his youth for a pure religion
of nature.[28]

These and other nature writings in the first half of the twentieth century made up environmentalism's Old Testament, a heritage that opened
an individual way to nature. For its New Testament, environmentalism
required work that moved from the individual encounter with a world
beyond society to consider our species' responsibility for life on earth.
Rachel Carson roused the public, but *Silent Spring* entered the canon as
a largely unread classic. Aldo Leopold's *Sand County Almanac* became
the Gospel. Since environmentalists came from social backgrounds that
looked on enthusiasm as faintly vulgar, no one sold WWAD ("What Would
Aldo Do?") bracelets or limp-leather-bound editions of his work, and the
twenty-fifth anniversary printings of *Sand County* did not print key pas-
sages in red. They did quote the land ethic on posters and conference
announcements. Academics combed Leopold's files for manuscripts, collected his articles for reprinting, and even put together *The Essential Aldo
Leopold* with quotations arranged to show how his thought developed.
Leopold's Sand County farm and the Shack remained in family hands, but
people came on the property and whittled away for souvenirs (relics) the
stump of the tree Leopold wrote of cutting up for firewood in "Good Oak."

In addition to a literature, environmentalists adopted symbols of faith
ranging from icons to areas. Environmentalists put up posters with Ansel
Adams's views of Yosemite Valley or Eliot Porter's pictures of Glen
Canyon in the same way that ethnic Catholics put statues of the Virgin
on the front lawn—as declarations of faith and reminders of what was
important. They took on the already established cause of saving the redwoods, symbols of nature's majesty whose groves became shrines for meditation, adding to the genteel campaign to buy the green cathedrals for
the park system the aggressive tactics of direct action. People lay down
in front of bulldozers and camped out in trees to save them from the chain-
saw. In the Pacific Northwest, environmentalists made the northern
spotted owl a symbol of the complex of plants and animals that made the
ancient forests a holy cause. (Opponents as well as defenders seized on
the owl, and one company capitalized on this by selling—as a joke—boxes
of "Spotted Owl Helper.")

Species that could be seen as essential to ecosystems or an indication
of their health became totems. In the 1920s few people, even nature-lovers,

objected when ranchers and government trappers wiped out the last breeding wolves in the American West. In the 1970s environmentalists made the wolf a symbol of healthy land and its presence a wilderness benediction. Tourists flocked to wolf country and enthusiastically joined wolf-howling groups in the national parks (which worked because wolves responded to even bad imitation howls). Wolf kitsch (mugs, coasters, scarves, T-shirts, even cribbage boards) became a growth industry.

In the Rockies, the grizzly assumed a similar status. Its presence seemed a guarantee that food chains were complete and the chance to see one a vital addition to the wilderness experience. In the Pacific Northwest, salmon became a symbol of the region and of the country's health; people followed the dwindling runs, while airport gift shops and tourists traps stocked smoked (usually farm-raised) salmon. The boxes came decorated in a style suggesting Northwest Coast Indian carving, tying modern veneration to earlier cultures' reverence for the fish.[29]

Environmentalism needed symbols, but also the patterns and habits that led toward right conduct. These qualities of character could be called "virtues," for they were habits of the heart that led to right living, the sort of action theologians had in mind in defining virtues as "practical habits of the intellect." Any way of living called for certain ones, and each system, secular or conventionally religious, drew from the culture's common store, emphasizing some, passing lightly over others, ignoring inappropriate ones or redefining them. Even in the early years, environmentalists emphasized some special virtues. They honored courage, particularly physical courage in defending nature (such as lying down in front of a bulldozer), but placed greatest emphasis on humility—taking that in the older, formal definition of a proper understanding of one's place in the world (Uriah Heep's self-abasement was not humility but pathology).[30] At the core of environmental sentiment lay the belief that humans, individually and as a species, must take their place as one species among many and not act as if they are the lords of the earth.

In calling for humility before nature, environmentalism extended a line of thought that went back to natural theology, but its call for us to "listen to the land" demanded a kind of listening that went beyond natural theology or nature appreciation's belief in the majesty of the world.[31] Environmentalism emphasized, as well as the insignificance of our species, our dependence on nature, and it demanded different kinds of knowledge about nature. Consider the popular bioregional quiz that first appeared in *CoEvolution Quarterly*.[32] It began by asking about our immediate surroundings—the phase of the moon, which direction was north, and the cycle of life during the year—went on to the catalogued knowledge of natural history and asked about events that shaped the land. Questions about the region's dominant plant associations required an understanding of ecological concepts and processes, those about earlier people and extinctions, an historical and anthropological understanding.[33] Doubtless, many people used the quiz to impress their friends, but it had a serious aim— to encourage people to learn about the land and their connections to it so they might understand who they were and what place they had in the world.

Where nature appreciation called for individuals to recognize their place in the world, environmentalism asked society and all humans to acknowledge that we were part of nature. Rachel Carson established that theme at the beginning of the environmental movement with her demand that we be humble before the forces that shaped us and her condemnation of "the control of nature" as a "phrase conceived in arrogance, born of the Neanderthal age of biology and philosophy, when it was supposed that nature exists for the convenience of man"—words her critics saw as an attack on the foundations of Western civilization.[34] *A Sand County Almanac* preached the same gospel. In the first section, Leopold spoke of individual joys and sorrows but in the later essays presenting the land ethic and the land community he emphasized humanity's place as one species among many. Declarations during the first years of the movement made the same point, and environmentalism steadily expanded its reach. The first pictures of earth from outer space became icons of the unity of life, and calls for international environmental protection relied on that view of earth as a fragile ark of life. The movement's integration of social justice with environmental concerns reinforced the point—we are all in this together.

Somewhat more tentatively, environmentalism grappled with a third aspect of humility: its own relation to the society. The moral righteousness that marked the counterculture encouraged the belief that reform positions were not only right but self-evidently right and that other interests could be ignored, and in the early years environmentalists sometimes proposed immediate, sweeping changes in the economy without much concern for what that meant for people's lives. A few environmentalists favored ecological authoritarianism, arguing that we had to bypass the slow and blundering processes of electoral democracy in favor of swift action (no one said just how environmentalists were to get the power to do this). That mood, never strong, soon faded, leaving only a supply of ammunition for the opposition.

Environmentalists increasingly turned to cooperation, moving from legislative alliances that called for no new tactics or ideas to coalitions and programs that did. Reintroducing wolves, for instance, required the consent and even the help of the people who lived on the land, and Defenders of Wildlife, working on the principle that environmentalists could not expect ranchers and farmers to pay for their principles, developed in the late 1980s a program to pay ranchers for animals the wolves killed. A few years later mainstream environmental groups, after initial hesitation, expanded their interest in the environment and their definition of it by joining inner-city residents to work against contamination and pollution in the cities. The most important, but least visible, cooperation was local. Many people, often informally and temporarily, made alliances and formed groups to deal with some environmental problem, in the process building a sense of community and developing their own skills in local politics. Some environmentalists (too many) continue to ignore other interests, but in the last twenty years much of the movement has reached out to other groups in society and recognized their needs.

Besides humility, environmentalism stressed hope, which it had overlooked in favor of optimism when it seemed the movement would soon triumph or we would all perish. As the industrial machine ground on, the environment deteriorated, and it became clear we would not solve our environmental problems quickly, optimism could not sustain commitment. Those who could not face the bleak prospect of a long battle ignored the news and quit the movement. Others held fast to optimism by believing that we had solved or were solving our problems. Gregg Easterbrook's *Moment on the Earth* (1995) argued (with more than 600 pages of text and citations) that there was already a "far-reaching shift toward the positive in environmental affairs." "Acid rain, pollution, radiation, overpopulation, and all the other looming disasters were not problems, not as bad as they seemed, or were being solved. "In the Western world the Age of Pollution is nearly over [and in] a very short time by nature's standards [our species] can become an agent for species preservation." Humanity was in a position to use its technology and intellect to "overthrow the basic restrictions of nature."[35] Six years later, Bjorn Lomborg's *Skeptical Environmentalist* (2001) made the same claims (with more footnotes), adding an attack on the good faith and honesty of the environmental movement. The *Wall Street Journal* and the *Economist* hailed Lomborg as the voice of truth; environmentalists and environmental scientists condemned him.[36]

Those who continued to fight for nature relied on hope, "that virtue by which we take responsibility for the future, not simply our future but the future of the world," which Gary Snyder described as the support that helped people take "the struggle on without the least hope of doing any good, to check the destruction of the interesting and necessary diversity of life on the planet so that the dance can go on a little better for a little longer."[37] In the 1980s, as environmentalism seemed stalled, more people consciously relied on hope. Bill McKibben's *Hope, Human and Wild* (1995) reflected that turn.[38] In an earlier book, *The End of Nature* (1989), McKibben had eloquently described the current of environmental despair that accompanied the realization that humans now affect every part of the earth.[39] In *Hope, Human and Wild*, he called us to go beyond despair, not by denial or optimism, but by hope and work. The specter of ecological disaster, he said, would not hold us to the environmental cause: "To spur us on we need hope as well—we need a vision of recovery, of renewal, of resurgence." Believing that the scientists are wrong or that we will somehow muddle through is "not hope—that's wishing. Real hope implies willingness to change." McKibben began with the situation around his home in the Northeast. Centuries of human occupation had largely destroyed natural ecosystems, but even here the land was recovering something of its wildness. It was giving us "a second chance. . . . The world, conceivably, will meet us halfway; the alternative to Eden is not damnation." He then gave examples of how other people, living in societies poorer and less industrialized than ours, have managed environmental problems. In the Brazilian city of Curitiba, experts grounded in the political and economic realities of urban life built a transportation system that accommodated people rather than cars, arranged space in ways that encouraged community rather than separateness, and made a livable city in a poor country. In the Indian state of Kerala, economic reforms and social programs slowed population growth and brought prosperity, giving "the lie to the idea that *only* endless economic growth can produce decent human lives." Back home his county's successful fight against a new landfill that would bring in trash from a distant city suggested that small successes in satisfying local needs would allow us to move away from the global economy and "toward something else—a community economy that over time assumes more responsibility for its own needs" and was more attuned to the land.[40]

McKibben's vision, though, rested more on rhetoric than reality. His examples of reform proved only that some things could work. He spoke of a local community and a "new economics and new politics [that would move toward] a civilization living closer to the limits set by its place, one that begins to solve not only its local problems but does less damage" to the world. However, McKibben's politics began and ended with town-meeting democracy, and while the town saved its land, the city's garbage would still go somewhere; McKibben not only could not say where it should go, he had no way to make the decision. He spoke of hope as something distinct from optimism, but did not clearly distinguish between them. His "vision of recovery, of renewal, of resurgence" that would help us face an uncertain future consisted of examples of good management and appropriate technology, and he admitted we needed more than that. A "community that made environmental sense," he admitted, "*would not have all the things that we have today.* . . . *It would be poorer,*" but he could only say it would be richer in (unspecified) personal and communal ways.[41]

To develop into a "community that made environmental sense," we had to decide what we really wanted and what we should do to satisfy our wants—but McKibben had no way to address that question. "Ultimately, along with all the questions about global economies and environmental imperatives, this is also a problem of desire, of what we're going to demand from our own lives." We had to confront the human desire for more, the demands that supported consumer culture, because, while our desires had no limits, the world did. McKibben did not deny the human drive for more, only that that drive was all or even primary. We "built our economy and society around one particular set of instincts and ignored the others, he said."

We had to turn to those others and cultivate them, learn to accept real limits, and pull back from our restless demands to change and keep changing things. Solving the "problem of desire" required a change of heart and a change of life. We had to "learn to take our wealth from the natural glory around us, to recalibrate desire so that we're satisfied by the sugar maple and not by the shopping mall." He did not or could not describe the satisfactions the maple offered, though, and fell back from hope onto fear. Change was not a matter of choice; we had no precedents and no outlines for such a change, "but do it we must; every signal from the natural world tells us that we now loom too large."[42]

For a variety of reasons, the environmental movement shared McKibben's confusion. It addressed ultimate, religious questions without the language or concepts of religion or even those from the academic study of religion. To complicate things it crossed many of the boundaries the culture enforced between reason and emotion, relying on science to describe nature but seeking spiritual experience and emotional connections to the world, and a purpose for individuals and the species, accepting a material world and material explanations but explaining our relationship with the world in moral terms. Other secular creeds had these problems, but environmentalism faced them in a larger framework, being concerned not with society or triumphant humanity but with the course of life on earth. Its confusion had another source as well: the unprecedented pressure humans placed on the living systems of the world. Three hundred years ago, so few human changes were visible in many lands that Europeans could describe them as "untouched" or "pristine" or "wilderness." Little more than a hundred years ago the great temperate grasslands remained unplowed, and there seemed an abundance of timber. Now "wilderness" required protection if it was to stay that way. It would take time to build a new picture of the world, or even find language to discuss it.

Measured against their problems, environmentalists accomplished much. Their early activism established environmental views within American society and government, put into place programs to save nature's beauty and preserve human health, and began debate about building an economy that would preserve the environment. After the 1970s, they faced the more difficult task of finding values for living. Because that lay outside the established search for meaning in nature and beyond political reform, the confusion and contradictions so evident is the 1980s should not be seen as signs of collapse but as products of environmentalists' engagement with these deeper questions. Having reached the limits of accepted categories and conventional action, they had to find new paths, and that meant wandering and stumbling.

That brings us to the present and to the movement's current problems. By its own logic, environmentalism must succeed or die, change American society or face the death of some essential part of the human spirit— and possibly the death of much of our species. That requires, in the long run, that people have a conscious relation to nature that is important to their lives, for they will not work to save what they do not love or love what they do not know and have made part of their lives. The next, last, part of this book takes up the fundamental obstacles environmentalism faces in making nature central to Americans' lives, not in terms of tactics, strategies, or particular ideas but on the fundamental level of ultimate questions, the level where the movement draws its strength. At the end it asks how we might benefit from more consciously acknowledging the depth of our environmental commitment, bringing us back to the question we began with, the relation between environmentalism and our answers to ultimate questions.

6 / Conclusion: "Quo Vadis?"

The King James Version of the Bible translates "quo vadis?" as "Whither goest thou?" but there "What now?" might be better. At the start I asked you to suspend judgment and—as an academic proposition—view environmentalism as part of the human impulse toward religion. That perspective framed the discussion of environmentalism's roots and its main lines of development. Having reached the present, we return (now on the level of society and not the individual) to the ways that "religion," "science," and "nature" meet and mingle in our lives. The view from religion casts light on that and it suggests what tasks environmentalists face. We should also, though, consider what use that perspective may be to the movement. Environmentalism's future and the use of a religious perspective still require the past.

While the very small print in the mutual fund brochures rightly warns that "past performance is no guarantee of future results," past performance is all we have—all we ever have—and while it cannot predict with accuracy, it can suggest with fair certainty. For example, if you had asked my mother in 1953 what would be in my room in 2003, she could not have told you, but had you told her that a glance around my study in 2003 showed—besides a load of books—a grackle feather; a turtle shell; the skulls of a field mouse, rice rat, and mockingbird; two airplane models and a ship model; and a few small chunks of coal and water-smoothed basalt, she would not have been surprised.

It seems certain that environmentalism has a future, for it developed in response to real problems and deep human needs that still exist. It is hard to believe the magic of the market or the march of science will halt or repair the damage industrial society does to the world's ecosystems. Environmentalism also speaks to disturbances in the human heart that seem the hallmark of modern society. For two centuries, a long line of social critics have lamented the loss of community and even the staunchest conservatives and believers in Progress interrupt their hymns to the market, free enterprise, and technology for ritual dirges about the decline of traditional virtues, the passing of small town life in which people really knew each other, and the rise of an anonymous society.

Environmentalism addresses the alienation in modern society in a modern context, finding the sacred in a material world and a way of personal knowledge and engagement with the world on the basis of objective knowledge. Grounded in science but also in Romanticism, refusing to choose between intellect and emotion, environmentalisms moves—more easily than either scientific materialism or established supernatural religions—across the borders between faith and knowledge, ignorance and mystery, offering material explanations but looking beyond facts toward meaning. It speaks of giving to the world as well as taking and of a purpose more active and important than the triumphant March of Humanity to the Stars by way of forays to the mall.

If environmentalism's survival seems assured, its victory does not. In forty years it has changed but not transformed American society, and its established tactics and programs seem unlikely to do much more than they have. Here the religious perspective offers what the view from political and social reform does not: a way to build a new life from within the old. If environmentalists seek a way of life more attractive than consumer society's, one that will sustain them in their cause and bring others to turn from the mall to the maple, then they need to look at the example of religion, for religions commonly live in one reality while preparing for another.

This chapter deals with three aspects of that work. It outlines the difficulties environmentalists face living in a society that opposes their core values but appropriates them, moves to the larger question of defining core American values in ecological terms, then takes up the possibilities and problems of forming an environmental myth. All look difficult. Envi-

ronmentalists must live within and even use consumer society while trying to change it, find what it means to be individual in an interdependent world, what freedom means aside from the chance to take as much
as we can from nature, and work out a vision of living with as well as from
nature. Finally, we turn to what environmentalists might gain from a religious perspective. What would be the advantages of more consciously seeing environmentalism as a way of accepting the universe?

LIVING IN THE OLD WORLD, PREPARING FOR THE NEW

Once the fervor of the early years faded, environmentalists faced the task
of living in a society that did not live by their principles. They could not
abandon industrial civilization to its wicked ways and retreat to the woods
(for the civilization was cutting the woods down) and in making a life
and furthering their cause the anti-environmentalists were the least of their
problems. Open attacks could be met, and they helped unify and inspire
people. James Watt's opposition to environmentalism, for example
brought a rush of people to the Sierra Club in the early 1980s. The more
subtle problems lay within, for even when they seemed to reject the society, environmentalists often incorporated its values.

Ecotopia, for example, believed as firmly as free-market Republicans
that science showed us the way, and (more subtly) that humans were
autonomous individuals. Ecotopians created local communities by voluntary association and nowhere did self-realization clash seriously with
group needs or did Pacific Northwest ecosystems limit their society. Green
consumerism rested on the belief that we could buy our way to responsible living, reshape the society by individual choices made in the market. The *Whole Earth Catalog* united human power over nature with
environmental living, proclaiming that if only people chose the right
books, tools, and philosophies, they could transform the culture—and
become as gods. Bioregionalism built on an accepted tradition of utopian
communities that included the view of completely autonomous individuals. Consciously, environmentalists rejected conventional values but used
them to think with.

Consumer society itself posed a formidable danger to environmental ideals. Its ceaseless demand for novelty made every popular cause grist for the advertisers' mill, which turned values into fashionable stands, and the more deeply felt the value and the greater the public interest, the more advertisers used it. Advertising presented the environment as an amenity, a part of the consumer life, not ultimate reality or sacred space, but part of a consumer lifestyle.

Commercials sold SUVs and ATVs as magic carpets to wilderness. Computer companies pictured people with laptops in mountains areas John Muir would have treasured and turned Muir's "fountains of life" and "God's Temples" into a backdrop for Mobil Oil or Georgia Pacific Lumber Company spreadsheets. The Nature Store turned environmental responsibility into a market niche; nature theme parks made it entertainment. People touched orcas and watched trainers put them through elaborate routines while announcers told them that watching the show proved their concern for nature. A real-estate developer sold a new subdivision under a quotation from Aldo Leopold: "Conservation is a state of harmony between man and the land" (Hypocrisy is the homage that vice pays to virtue).[1]

Because environmentalists wanted to save nature's systems, and not just parts of nature, they faced problems earlier movements for nature had managed to avoid. Without hypocrisy Muir could use mass-produced magazines to call people to the wild and enlist California's elite and the Southern Pacific Railroad on behalf of a national park in Yosemite because he aimed only to preserve areas of nature, and it seemed possible to separate nature and development. Environmentalists could not take that way out, for the world was limited, all things connected, and all nature under siege. Rejecting growth as a solution, calling for sacrifice, and warning of limits, they stood against common American assumptions about the good life. They had, however, to work in that society, and their tactics suggest that Thoreau was right when he said that ways of getting money without exception lead downward. Memberships never provide environmental organizations with enough money for all they need to do, but environmentalists view donors with suspicion. Even if they did not believe that behind every great fortune was a great crime, they suspected that great exploitation of nature was. Accepting money seemed a way of allowing donors to buy an air of environmental virtue or, worse, corrupt the cause.

They could raise money by using consumer society against itself, but that produced more contradictions. The catalog of the National Wildlife Federation (NWF), entitled "Sharing Nature," featured goods "designed exclusively for us by renowned wildlife artists," and assured readers that "every single dollar of profit" from sweaters, wineglasses, and stuffed toys went to "continue our work to help protect and preserve our wildlife, natural resources, and the earth's environment."[2] So, presumably, do profits from the NWF credit cards in my wallet, issued by some enormous bank. When indigenous people gather nuts from the rainforest or take specimens to be tested for useful drugs, their work puts money into traditional economies and staves off the loggers, but it also puts the forest on the market, implying that if logging became more profitable than nut-gathering, the forest could be cut down.[3] Ecotourism kept areas "pristine" but reinforced the view of wild nature as an amenity for the upper crust rather than the ground of human life.

Critics on the outside accused environmentalists of hypocrisy or lack of business sense; those within spoke of betrayal and called for action on principle. Defenders of these practices usually offered some version of the old adage that occasionally the Devil had to be beaten with his own tail. Living a pure life in an impure would is a classic religious problem, arising out of the difference between the world as it is and as it should be; environmentalists, unfamiliar with terms like "laxity" and "Puritan rigor," usually borrow their vocabulary from politics and the counterculture and talk of "sell-outs" and "crazies."

NEW TERMS FOR NEW THOUGHTS

Public defense of environmental practices brought out a more troubling problem—the way the terms and assumptions of public discussion undermined environmental values. As Aldo Leopold—among others— pointed out, American society measured value by dollars, but most parts of the land community had no obvious economic use. To appeal to legislators environmentalists argued that endangered species should be protected because a frog or flowering plant in the rainforest might some day yield a cure for cancer, and they pointed to the dollars that would flow into the local economy when the old-growth forest around it received official blessing as wilderness. Their arguments, as many recognized, worked against the cause even when they worked, but while environmentalists recognized the contradictions, they did not usually see how deep they went. Without environmentalists fully realizing it, the movement challenged accepted definitions of basic American concepts like freedom and conventional ideas of how the individual and the community were related, and in this hidden clash, more than in discussions of policy or what constituted a true biocentric perspective, lay the movement's truly radical break with the conventional wisdom.

Environmentalism challenged both what we knew and how we came to know it. Our ordinary view emphasized separate things and dismissed relationships and regard for human involvement in nature as matters of choice. Ecology took the opposite point of view, making relationships primary to the point of defining organisms by their interactions with others. That perspective turned our definitions of individuals and community upside down. Rather than being fully autonomous beings that used the world outside them as they wished, individuals seemed in some essential way to depend on the world. The community ceased to be an amenity individuals formed by choice and became instead a necessity that existed before them and in some sense made them what they were. Beyond that, environmentalism's commitment to personal understanding and involvement demanded a definition of knowledge beyond the "objective" reproducible information the society accepted. Its belief that nature had value required a new scale of values. Industrial society's destruction of natural systems had to count as more than just personal feelings or even economic value thrown away.

Environmentalists had to change society's definitions, then, not only of what the world was but of what was valuable in it. That did not call for anti-environmental rhetoric, substituting feelings for reason. It did mean giving nature (as well as the market) value in society and people's lives, accepting the search for a place in the land as a way to live as valid as trading stock, and weighing the social value of that community as well as individual greed. That involved changing, and enlarging, the American faith. Its central propositions, like the fully autonomous individual or the definition of freedom as the chance to shape the world, seemed rational and scientific, but were, in fact, a creed. Just because millions believed in Progress, the market, the unconquerable spirit of mankind, and endless economic growth, that did not make them accurate descriptions of human motivation or the world's ways. That humans were rational by nature and by nature desired more things came from an eighteenth-century psychology that assumed that certain human drives (drives eighteenth-century English and Scottish intellectuals found in their own lives) were basic, and all humans as basically alike.

Environmentalism's conflicts with the accepted view of freedom suggested how deep that idea went and how deeply the new values challenged it. At first, environmental legislation seemed the sort of necessary restriction on individual freedom that everyone could accept. Even the strongest defenders of a laissez-faire economy agreed that people could not use their property to damage others, and since pesticide residues reached everyone, not merely those who lived near the factories, and some manufacturing wastes remained dangerous for generations, public health and safety required regulations on these materials. Manufacturers could not openly argue that profits were more important. They usually defended their operations by arguing that science had not "proved" the chemicals they made caused any damage or that they could change their ways and keep making the materials, but in the early years generally lost the argument. The nation banned DDT and several other pesticides and regulated many more.

The issue of freedom emerged more clearly as environmental laws affected land use, for property in land had not only, historically, been a private matter, but it had been identified with economic opportunity, the very symbol of American freedom. The inland West saw the most intense conflict, for it held most of the nation's public lands and its economy depended heavily on them. Ranchers who had for generations grazed cattle on public lands resisted reductions in their herds, larger fees, or changes in grazing practices. Miners fought new laws that restricted exploration and claims, and loggers and lumber companies protested new timber policies that looked for environmental protection rather than timber production. They all spoke about the economy but more fervently of freedom, defined as people's ability to do what they pleased on land they owned or used. James Watt, the secretary of the interior under President Ronald Reagan, became the environmental movement's Great Satan in the 1980s for his advocacy of this position. Emotionally loaded programs like the reintroduction of wolves, first into the northern Rocky Mountains and then into the Southwest, raised tempers and rhetoric even more.

In the 1980s the opposition to environmental legislation rallied around the banner of "wise use." In theory it accepted regulation and looked at the costs and benefits of policies, but in practice it identified "freedom" with established economic arrangements and environmentalism with big government or socialism (the distinction was not always clear). It accused environmentalists of "locking up" land for the recreation of wealthy people from the cities, denying ordinary people their chance at the American dream. The charge carried considerable weight in the West and the culture at large because Americans identified freedom with the chance to prosper, and prosperity with access to real property and the chance to use it as they wished.

While that might not have been true in the contemporary West the belief had a long backing in American history. The land laws of the early republic consciously aimed at a nation of independent farmers working their own land, which in a preindustrial society was the only source of production. That trend culminated in the Homestead Act of 1862, which promised a free farm to anyone who would work it. Even when farming ceased to be Americans' ordinary occupation, legislators continued to speak of freedom in terms of opportunity to advance oneself. They spoke of encouraging small businesses and breaking up monopolies. Conservation measures passed at the end of the nineteenth century regulated use of natural resources, but conservation agencies promised to allow small companies access to timber and other goods. Preservation only asked that small areas of nature be set aside for nature, and even these yielded economic benefits.

Environmentalists needed a new conception of freedom that took account of communities and common goods, but they found little to build on in their own traditions. From Thoreau at Walden Pond to the modern wilderness journey, Americans went, as individuals, to nature. Environmental activists emphasized individual action, bioregionalism, the individual life formed on the land. Bioregionalists did look forward to a new community on the land, but that would come from the accumulation of individual choices. In rethinking community and freedom, environmentalists had a hidden ally—the culture's own disputes about freedom. Although the strength of convention made it seem that freedom had one accepted meaning, people always argued about what it meant.

Drawing on conceptions of freedom that went back to the foundations of Western civilization, Americans gave the word almost as many meanings as they did "nature." They defined it as the power to do as you wished, the result of subduing your passions to surrender to the will of God, or life according to the dictates of Nature. The generation that established the nation also established freedom as self-assertion by autonomous individuals, but they did more than provide documents (the Declaration of Independence and the Constitution) for their descendants to interpret. They gave them concepts to reshape and use. Every American reform and reformer promised to extend freedom to still-oppressed groups or give more of it to those who already enjoyed its blessings, and in every discussion of social policy both sides appealed to it. Businessmen argued that labor unions destroyed the worker's freedom to enter into contracts for his labor, while union organizers spoke about workers' freedom to associate for their own benefit. Advocates of woman's suffrage cited the Declaration of Independence; their opponents appealed for a social sphere free from government interference.[4]

Environmentalists, though, faced an exceptional challenge, for the ecological view, which put communities at the center and emphasized necessity rather than choice, made it difficult to appeal to or to get past the old verities. They had some room to maneuver, however, for not even the scientists, on whose work environmentalists relied, agreed on the importance of communities. They were central to Frederic Clements's theory of plant succession, which dominated American plant ecology for two generations, but ecosystem ecology, which developed in the 1940s, rejected Clements's view of associations, not individual plants, as fundamental units and paid less attention to balance and stability in ecosystems. In the 1980s its research emphasized individuals, competition, and instability. Still, no ecologist denied that organisms existed within systems and functioned within limits set by the environment.

Environmentalists did not need to reject the idea of the individual; they needed to strike a new balance between the individual and the community, one that took into account both human freedom and human dependence. Humans certainly depended on nature, even when they escaped any single biological niche by living in and drawing from many different ecosystems. On the other hand, no environmental way of living took people fully back into an ecosystem. Even the most committed bioregionalists remained part of human culture as well as biological nature.

Defining freedom under those conditions required environmentalists to come to terms with the element of necessity in human lives. American society strongly emphasized individual rights, but Western civilization as a whole offered other views. The balance between individual and community has formed the central question in Western political thought since the days of Socrates and Plato. The problem, however, was less theory than its application in our lives, and people, particularly in the early years, tried many new paths.[5] One account of efforts to develop a bioregional life in northern California, Peter Berg's *Reinhabiting a Separate Country,* included "tape-recorded stories and interviews, poetry, essays, prose, drawings and photographs." It described "reinhabitants exploring ideas and activities for living-in-place" with everything from Native American wisdom to accounts by "natural scientists defining and describing the place."[6]

Lacking a vocabulary, concepts, or even a line of action adapted to their project, bioregionalists tried what they could, discarded what failed, and held on to what worked. They discarded dreams of total independence, easy communication with the land, and perfect harmony for the rigors of a life-long search. Wendell Berry's description of one such life, in *Jayber Crow* (2000), one of his novels about the fictional town of Port William, Kentucky, suggested some of the dimensions of that kind of life. It told the life of Jayber Crow, who left Port William as a child, came back as an adult, became the town barber, and at the crux of the novel made a personal vow to live in the town and the area around it, using its resources and making himself fully part of his community. He stopped driving and got rid of his car, a decision that "in itself was an important stage in [his] journey . . . [for it forced him to] make do with the company that came," to build a life around the local society.[7] Berry showed in the novel the richness of a life lived with others and the joys of a man who filled a niche in the town and devoted himself to it.

Though Jayber lived a secular life in a secular community, his vow paralleled that of Benedictine monks, who pledge to remain in the community they join, to make their lives and find God within its walls with the others God has called there. Like them, Jayber made his life by living it rather than thinking about it—only in living it did he discover what his decision meant. Strange as it seemed to modern sensibilities, his choice emphasized freedom, not freedom as moderns saw it—access to power or the ability to direct one's life—but freedom as classical civilization and Christianity saw it, action in accord with the realities of the world. Neither bioregionalists nor the larger environmental community spoke consciously of that kind of freedom, but some explored it, and their work may be the start of a new understanding of freedom in the context of environmental action and a sustainable society.

AN ENVIRONMENTAL MYTH

Environmentalism faced the long term because it had to transform American society and so far has only affected it. Environmental campaigns established laws and policies and made concepts like humans' obligations to the land and the nonmaterial benefits of ecosystem preservation familiar enough for newspapers to use them without explanations or quotation marks. But while people told public-opinion pollsters they thought pollution was a major problem, they bought bigger houses and bigger SUVs and continued to believe that endless economic growth was possible and necessary.

Against that wavering sentiment, environmentalists needed a coherent vision, a description of the world's realities and a life lived with nature compelling enough to hold people to the cause through a lifetime of effort.

They needed, in short, an environmental myth, the kind of vision that religions, even secular ones, used to show their beliefs. Marxism, for example, organized all human history around the class struggle and gave everyone a place and purpose within that great tale. Marxists disagreed, often bitterly, about the "correct" Marxist interpretation of any historical phenomenon, while agreeing it applied to every part of life and to everyone. America had its own Dream, less intellectually organized (the rejection of ideology formed an important tenet of the faith), but with the support of the very material blessings of consumer goods. In it the United States, the vanguard of democracy and capitalism, molded the inert matter of the world and led the rest of humanity into a blessed future. The striving entrepreneur and the careful consumer moved the nation forward, and all had the hope of even greater opportunities for the next generation.

Environmentalists could not, as believers in supernatural faiths could, appeal to certainties beyond our understanding and a place beyond this world where all tears would be wiped away and all injustices righted, and they could not offer material blessings comparable to consumer society's goods and gadgets. Despite searching beyond knowledge and reason for meaning, they could not rest their case on explicitly mystical or even spiritual ideas. Worse, their case required them to hold in tension the apparently incompatible perspectives of reason and emotion. If our ties to nature were only physical, and our love of nature a reflex left over from evolution, our ties to nature became just another atavistic impulse that therapy or a few generations in the world-city would cure. If our connections to nature were only emotional and spiritual, environmentalism and the preservation of nature's diversity dwindled to the enthusiasms of another special interest group.

As it has in finding a new description of freedom, in making a myth environmentalism relies on turning existing currents in the culture in new directions. These helped even in its movement across the borders between faith and knowledge, mystery and understanding, the sacred and the profane. These appear opposed, but much of life involves holding apparently contradictory views, recognizing each as a true though partial perspective. People also want meaning in a material world, even if scientists tell them nature offers no moral lessons and philosophers say they cannot use what is to say what ought to be. People also want some cause worth working for, even worth dedicating their lives to. Environmentalism offers that, and does it in science's secular terms. Its myth, incomplete and controversial as it is, accounts for much of the movement's power to hold its believers and to recruit from a new generation.

To move beyond that incomplete description, environmentalists have to change their own perspective. They have not rejected the heritage of wilderness but are turning more consciously to a strain within American thought about nature that encouraged people to nourish their lives in nature by finding nature where they lived. That went back to Thoreau, who wished to speak "a word for Nature, for absolute freedom and wildness as contrasted with a freedom and culture merely civil," and proclaimed that in wildness the preservation of the world but found nature's essential qualities in Concord.[8] Cronon pointed to that search when he asked us to look at the tree in the garden as well as the tree in the wilderness. So did Pollan, urging us to find nature's life through gardening. So did the campaigns for wild urban and school parks, restoring our backyards, and planting native shrubs. Bioregionalism followed as it looked for ties between people and the land that would unite home, field, and wilderness, work and play.[9]

Daniel Botkin continued that line in *No Man's Garden* (2001). Facing the "question of whether modern society can achieve a connection between the human spirit and nature, and between civilization and nature, that builds on the ideas and tools of Western civilization," he turned to Thoreau for help in working out our "connection to nature, both physical and spiritual," and the ties between civilization and the environment. He used Thoreau because he believed Thoreau's path went between deep ecology, which seemed to favor the end of civilization to preserve the biosphere, and the "wise use" movement, which saw environmentalism as a conspiracy against humans.[10] Anti-environmentalists saw no value in wilderness, but arguments for wilderness "tend to take on a puritanical tone; that nature is good and people have been evil, that people must be kept from the sin of destroying nature, and kept separate from it."[11]

Because Thoreau valued both civilization and nature and focused on the relationship between humans and nature, his life provides "a metaphor for the search for a path to nature knowledge and a resolution of the questions inherent in humanity's relation with the rest of the natural world." It points toward "a land *and* people ethic. It has two components. One is the civilization-ecology ethic; the other is the individual-nature ethic. . . . Neither is possible without love, understanding, and faith—what I refer to as spiritual or religious qualities."[12] Those qualities were essential. Environmentalists used utilitarian arguments because "there is a sense that nonmaterialistic arguments will not 'sell,' will fail to hold up against economic arguments. . . . But in the retreat to materialistic justification, there is an implicit retreat *from* a belief in the power of ideas and the power of human passions."[13] Environmentalism had to assert that power in the face of a culture that saw the market as the only arbiter and against a strain within the movement that rejected humans.

Botkin steered between the extremes of deep ecology and wise use by refusing to see humans as either part of nature or apart from nature. He accepted them as at once biological beings immersed in the world but also creatures who could, mentally, stand apart from the world and, at least in the short run, live without reference to nature's limits. Many people found that stand as strange as religious language, for they believed, along with many Western philosophers in the last two thousand years, that at least in principle all questions had single, simple answers and that humans could find them. Environmentalists who insisted on a sharp separation between nature and culture shared that view, but others accepted blurred boundaries and partial truths as they did in ordinary life. People, for example, thought of themselves as individuals and cherished their freedom without denying that they belonged to families and nations, and they routinely applied different sets of standards to different areas of life (even academic economists did not measure family love by cash value).

In the last forty years environmentalism has moved toward an understanding of the good environmental life and its goals. Progress did not come from philosophy and reform programs. Instead of guiding people in their lives, deep ecology, social ecology, ecofeminism, and other critiques showed how environmentalism and other programs were related.[14] Progress came from action—everything from recycling to spiking trees— carried out in places as different as wilderness and backyard, some of it useful, some (inevitably) mistaken or wrong-headed. People explored an environmental life in the only ways they could. Using nature appreciation's language and ideas, environmentalists eloquently described spiritual journeys and paths to an individual life on an individual patch of land. They established a guide to self-realization in nature, a nature that included humans and took account of human society.

What remained unclear was how we were to move from an economy built on ecosystem destruction to one that restored and nourished nature, what that would look like, and what satisfactions it would yield for people, the majority of whom were not farmers or environmental thinkers and activists. Clashes over old growth in the Pacific Northwest revealed the tensions between environmentalists and people making their living directly from nature in obvious ways, such as in loggers' bumper stickers that said: "Are you an environmentalist or do you work for a living?"[15]

SEEING ENVIRONMENTALISM AS A RELIGION

The loggers' bumper sticker was one sign among many that making an environmental myth calls for more than zeal, practice, and patience. It requires a fundamental view of the movement's aims and a clearer understanding of its challenge to basic values and beliefs. That takes us back to the beginning of this argument. If environmentalism strikes to the level of our basic beliefs about humans and their relation to the world, might environmentalists gain a better perspective by acknowledging their commitment as a religious act and their beliefs as a way of accepting the universe? The first question is: why don't they do that now? Environmentalists clearly see their cause as fundamental, link our destruction of nature with the great extinctions marking geological epochs, and even fear the end of evolution or of nature itself. They use religious tropes and language, admire primary cultures for integrating nature and religion, and borrow Native American ceremonies and Eastern traditions. Some spike trees, cut fences, burn down buildings, and a few die defending redwoods. However, even those who call environmentalism their religion put "religion" in quotes. They take, if only by implication, Devall and Sessions's position that religion means some ground of belief more basic than an environmental stand. From the other side, those committed to a conventional faith more commonly ask how their own traditions shed light on the environmental vision rather than what environmental knowledge says about their (conventionally defined) faith. What E. F. Schumacher called the general modern ignorance of the language and concepts of religion plays a part. Environmentalists do not think of their commitment as religious because the idea lies outside their customary lines of thought. The society's common view of religion as blind faith, detached from reason and even reality, a perspective well-rooted in the social class from which environmentalists commonly come, makes the idea of accepting something as a faith not only odd but somewhat distasteful to most environmentalists. Closer to home, admitting the depth of their commitment would require environmentalists more clearly to face the extent to which humans have destroyed the biological heritage of our planet and how much more is at risk. If half the scientists are right about half the things they are predicting, our world will change dramatically and permanently unless we as a species change our ways and do it soon. The death of nature and the loss of a world beyond are literally fear-full prospects, made more so by the obvious power of industrial society and the lack of obvious solutions. Statistics and case studies distance environmentalists (and antienvironmentalists) from the full impact of a possible, horrible future. Fully accepting commitment to wild nature as a fundamental part of our beliefs removes the armor of "reason" and "data."

Do I exaggerate environmentalists' fears? I think not. Nature writing, which once spoke of the death of species or the destruction of a familiar landscape near home, is now concerned with the loss of ecosystems and the diversity of life itself. The fear of losing nature appears in *National Geographic* and the daily paper as well as *Sierra* and *National Wildlife* in articles recounting the contamination of the ocean, global climate change, and the strategy of triage to preserve our planet's hotspots of biodiversity. An undertone of horror at environmental loss haunts even antienvironmentalist literature. Believers in the market, triumphant human reason, endless Progress, and the conquest of the universe overflow with optimism, but hymns to human progress and an improving world, books like Easterbrook's *Moment on the Earth* and Lomborg's *Skeptical Environmentalist,* protest a bit too much.

Anti-environmentalists marshal endless statistics and case studies, speak of new technologies and scientific advances, and use footnotes like a graduate student facing a hostile committee. Yes, they say, everything *is* going to be all right. There really *is* nothing to worry about. We know what we are doing and can deal with any problem that arises. They pile up statistics, graphs, and citations as if these would bury the spectacle of vanished forests, dead species, and leaking waste dumps, as if optimism would erase the knowledge that nature's wonders are dwindling from the impact of 6,000,000,000 people and their machines. If those who find no value but money in nature are so determined to avoid looking at environmental decay, can we wonder that those who love the world of nature with a passion find it difficult to admit how much it means to them?

Acknowledging environmentalism as an ultimate commitment raises the stakes and the prospect of pain, but what advantages might come from environmentalists' taking the perspective of religion? Perspective, it should be said, does not mean church or creed. The "High Priest of the Muir" and the "True Voice of the Tree" ought to stay in science fiction.[16] Accepting environmentalism as part of our understanding of our place in the world would, by emphasizing how serious an issue we face, show how truly radical a response it demands. Besides showing how far the movement goes beyond reform, seeing environmentalism in religious terms would focus discussions about what ought to be done, help environmentalists confront their opponents, provide a way to make common cause with outside groups, and allow environmentalists to appreciate more clearly the roots of their own movement.

The religious point of view shows the conventional wisdom for what it is, not a "rational" analysis of the human situation, but a faith resting on beliefs about people and their relation to the world. It suggests parallels between environmentalism and conventional religious beliefs, which, since denominations now take environmentalism and environmental problems seriously, would help the movement communicate with these potential allies. It would also help environmentalists learn from the struggles conventional faiths have had with science (a problem their movement faces) and understand themselves by seeing how much they have built on conventional religion. Finally, a religious point of view would help environmentalists move beyond optimism and short-range goals, for religion, in general, has faced that task for the last few thousand years.

Seeing environmentalism in terms of faith can help environmentalists meet their opponents on level ground rather than, as is now too often the case, from a position in which anti-environmental arguments appear as the product of reason, part of the way the world is, and environmental arguments as the emotional outbursts, for anti-environmentalism appears, from that point of view, as a faith. Just as the Puritans supposedly spent Sundays contemplating the goodness of God and the damnation of infants, defenders of the status quo gaze in rapture on the goodness of the market and the gross domestic product, which distributes benefits to all who strive—for they are the virtuous. They look on the market as the Hand of God, endless economic growth as the path to the Earthly Paradise, and the conquest of nature, human destiny.

Because society accepts these propositions as self-evident truths and rational descriptions of the universe, other views—ecological interdependence as the Hand of God, a sustainable economy as the path to the Earthly Paradise, and living as citizens of the biotic community as human destiny—look like emotional arguments, unworthy of consideration. So anti-environmentalists speak of Progress, the Questing Spirit of Mankind, and the American Dream (emotionally loaded terms) while appearing to argue in rational terms and on neutral ground. Framing public debate as the clash of competing beliefs would allow environmentalists to engage orthodoxy more deeply by showing it as a faith built on eighteenth-century realities and nineteenth-century history, one that fails to describe a twenty-first-century world where humans displace nature's systems on a global scale.

Religious terms would help environmentalists deal with conventional faiths, with which they are involved in several ways. Churches now take the movement seriously, something they did not do until the 1980s; when Wendell Berry began his bioregional experiment in a Christian framework, few supported or were even interested in it, and debate about religion and the environment hardly went beyond discussing Lynn White's "Christianity and the Ecologic Crisis" or parsing God's command in Genesis for humans to take dominion over the earth. As environmental problems multiplied, affecting more people in more ways, theologians and ordinary believers asked what their faiths said about moral responsibility to the land and what place Creation had in salvation history, and from questions proceeded to action.[17] Local churches addressed problems around them, and denominations established offices and programs.[18] The American Catholic bishops, for example, applied the Church's social teachings about the sacramental value of community and the concept of stewardship to treatment of the environment, and individual bishops and regional conferences issued pastoral letters—formal expressions of reflections on particular problems. The bishops of the Pacific Northwest published one letter, for example, on human responsibilities to the Columbia River and the life that depended on it.

Church members worked environmental studies and ideas into their spiritual lives and practices. *Spirit in Nature: Teaching Judaism and Ecology on the Trail* included exercises and lessons for children as young as seven that brought Judaism to bear on experience in nature and showed how experience in nature enriched Jewish spiritual life.[19] Rather than seeing this as a departure from tradition, the authors believed an "awareness of the complexity of the human relationship to the natural environment has figured prominently in Judaism's literary and legal landscapes." The blessings observant Jews utter, it pointed out, include many relating to the natural world—one to be said on encountering "shooting stars, earthquakes, lightning, thunder and storms," others upon seeing "natural wonders," "trees in blossom for the first time in a season," or "the rainbow."[20] There was a blessing "before eating fruit" and another "over rain and over good news" (a good combination for a people of the desert).[21] *We Are Home*, written from a Christian perspective, explained "why the health of the environment is a spiritual issue" and discussed homemaking and home finances by putting Schumacher's idea of the meta-economic foundations of economics in kitchen terms.[22] Scientists recognized the change in religious attitudes. "Professional scientists today," physicist Freeman Dyson observed, "live under a taboo against mixing science and religion," but in 1990 a group assembled in Moscow for a conference issued "an appeal for joint commitment in science and religion."[23] Environmentalists should do no less.

Besides using it to work with others, environmentalists could use a religious perspective to understand themselves. Ever since Emerson, Americans who failed to find God in church took terms and perspectives from Christian theology into their search for ecstatic experiences in nature. Environmentalism's rhetorical strategies, points of view, and ways of thought remain embedded in this evangelical Protestant heritage, which forms the unacknowledged ground of many environmental attitudes and arguments.[24] Bioregionalism built on the American Protestant traditions of utopian communities and individual spiritual searches beyond society. Green consumerism found its focus on individual action and individual salvation at the same source.

Religion, as Mark Stoll's analysis of the clash between Jeremy Bookchin and Dave Foreman suggests, influenced even those who had no formal religious faith. Bookchin, raised as a Jew, linked environmental action to social reform and criticized deep ecology and wilderness advocates for neglecting or even dismissing human beings. Foreman, from an evangelical Protestant background, looked at humans much as the Puritan divine Jonathan Edwards had—as a disease upon the earth—and found redemption in a pristine world of nature. Their views, Stoll argued, grew out of their "different worlds of culture and value. A community-centered, anti-individualist, tightly reasoned plan for utopia collided with a nature-centered, strongly individualist, evangelical message of salvation . . . [there was] a gap in the moral assumptions of different branches of environmentalism, a gap with deep historic, cultural, and theological roots."[25]

Environmentalism could learn from conventional faiths' experiences with science, for like them it needs to take account of that knowledge without fully accepting science's authority or the views often associated with it. When the movement began, environmentalists relied on ecology's view of the world to justify new policies and programs. They also accepted the dominant picture of nature, drawn from ecosystem ecology, which stressed balance and interdependence and seemed to show ecosystems driven by cooperation as much as competition. That view led Barry Commoner to proclaim in *The Closing Circle* (1971) that "Nature knows best." Environmentalists said amen, and a decade later many embraced James Lovelock's Gaia hypothesis, which suggested the world's biological processes evolved to favor life and the conditions that made for life and formed a self-regulatory whole.[26]

By the 1980s, though, ecological ideas did not so clearly fit environmental ones. The scientists had turned from the questions ecosystem theory posed to others, some raised by the answers ecosystem studies provided. Their work, focused on individuals rather than communities and emphasizing competition and chaos more than cooperation and harmony, produced a picture of the world in which natural systems were not stable and balanced, but easily disturbed and subject to rapid, irreversible change even without human intervention. Rather than cooperation, competition seemed the dominant force. That made it difficult to use science to justify letting nature alone or to attack human interventions in nature's systems. Environmentalists met the situation with more finesse than the fundamentalists battling Lyell and Darwin, but denouncing scientific apostates for adulterating the pure milk of science with the poison of free-market capitalism or with neglecting the work of the masters hardly met the case.[27]

Christianity and other traditions have faced this sort of problem for three hundred years, and the mainstream of Christianity and Judaism accepted science as true knowledge about the world, or at least about the material world and, therefore, as a limit on theological speculation. Since Truth was One, science could not conflict with revelation, and in its proper sphere (knowledge of the material world) its well-grounded conclusions should prevail. On the age of the earth and the arrangement of rocks, geology took precedence over Genesis. Theologians, though, continued to use science's knowledge to work out humans' place in Creation and Creation's role in salvation history.[28]

After the heady early years, when environmentalists found in science and technology either salvation or the cause of all our ills, they had to decide how far science could guide them. It gave information about the material world, the only data useful for public debate, but said nothing about the commitment to nature many environmentalists felt before any rational analysis, for it rested entirely on rational analysis and excluded emotion. Recognizing more consciously that science's formulas talked of truth but were not the entire truth—a view from theology—would allow environmentalists more easily to work out their allegiance to science and scientific information and their differences from its ethos and practice.

A religious perspective would also help environmentalists make sense of their spiritual practices, often borrowed from primary cultures and Eastern traditions, by placing them in the frame of a common human search for an understanding of humans' place in the universe. It would show that borrowing, for example, is common practice. Christianity not only took Greek philosophy as the basis for its early theology, it used symbols from earlier faiths, "baptized" local spirits as saints, and set up holidays to compete with pagan festivals. In that light, the wilderness journey's mixing of outdoor recreation, masculinity, and spiritual search seems less a problem than imaginative use of the culture's materials; its making Yosemite and the canyonlands into icons, the equivalent of establishing holy places; its totems and regional symbols parallels to the cults of local saints; and its naturalist cruises to the Galapagos with daily lectures, the modern equivalent of the medieval pilgrimage—part sacred journey, part vacation. The religious perspective would make sense of borrowings by focusing attention on the reasons environmentalists do it. They do not pray, but value the fundamentals which prayer encourages; do not have church on Sunday morning, but do seek community. Awareness of the reasons for spiritual practices would help environmentalists find useful equivalents of Wednesday-night prayer meetings, corn dances, sweat baths, or chanting.

The greatest value of a religious perspective may be in helping environmentalists face the long term. The movement began with crises and crusades—*Silent Spring,* the fight for wilderness, the campaign to ban DDT—and environmental organizations still rely on emergencies to rally the faithful and raise money. Facing emergencies, though, is not the same thing as facing problems. Demonstrations helped end the use of DDT but did not address the economic pressures that led farmers to use it in the first place, and unless the system changed they would—they had to—turn to other chemicals. Marching and waving banners outside a meeting of the World Trade Organization shows environmental principles, and direct action or selling nuts gathered by indigenous people saves some part of wild nature for the day, but they leave intact the global economic order that led to the problems. Green consumerism and bioregionalism fought consumer values without acknowledging environmentalists' participation in consumer society. The litany of warnings and string of emergencies eventually became counterproductive; people became suspicious of even well-founded alarms, and as most of the predicted disasters failed to happen in the form environmentalists predicted, anti-environmentalists used that to condemn the cause.

With the fading of the counterculture and the realization that industrial society was not going to collapse next year, environmentalists faced the challenge of living through a lifetime in the service of a cause that would not triumph soon, but must reshape the world if we all are to live in it. Religion (except for the creeds promising that the aliens were going to show up next year and save the righteous) lived with that situation, and environmentalists might consider what could be learned from the multitude of good and bad examples the churches present. Facing the long term requires environmentalists to accept, besides the prospect of failure and the need to continue working, the certainty of a diminished life. Even if we found a sustainable way to live tomorrow and everyone adopted it next week, we would still live in a world stripped of many biological treasures and without many of the undeniably good things the present world offers—from easy movement between the city and the wilderness to deeply held dreams of Progress.

Against those prospects despair and anger would not serve, nor would what someone once called the "true religions of America . . . optimism and denial."[29] Conventional faiths, though, nourished hope and encouraged action in the face of defeat—even inevitable death and the fading of all we built—and if they cannot serve as models, they can provide a perspective, for they share with environmentalism the belief that humans are limited creatures, living in a world that was not made for their convenience. In order to help more than a select group for a short time, environmentalism must confront the problem of long-term living. Seeing its commitment as a way of accepting the universe would be a start.

A religious perspective would not be a cure-all. It would not solve environmentalism's problems or even show a clear path to the future, for religions are less about solving problems than about confronting them. It would not change people, either. Even if everyone accepted environmentalism as their religion and acted on it, some would still plant water-hungry grass in their front yards, toss half-empty cans of pesticide in the trash headed for the landfill, and do from ignorance or laziness all sorts of other unsocial or un-environmental things. Certainly, open declarations of faith would create their own problems, allowing anti-environmentalists a greater chance to condemn the movement as a remnant of the "sixties," part of the counterculture and therefore out of touch with reality, or part of the New Age, out of touch with reality and opposed to true, Godly, American values.

Open declarations of faith, though, would be out of place, for while environmentalism is involved in religious questions and deals with ultimate commitments, it is not a "faith" in the sense of being a creed. What would not be out of place would be the perspective of religion, the view of environmentalism as a way of confronting the human situation in our world. That point of view would help sort out arguments, beliefs, and action, clear away mental underbrush, and allow people to explore issues now hidden because not seen. It would separate the trivial from the fundamental. While the relative importance of wilderness preservation and urban pollution control would remain as political issues to be debated, their relation to basic goals would be clearer. A religious perspective would not direct us to the garden or the wilderness; it would help us see what value each had. People could consciously deal in a unified fashion with issues they now confront piecemeal and unconsciously. Besides, it is never a bad thing to see your own position clearly.

In the preface I said the conscious beginning of this book lay in my reaction to a discussion of Cronon's "Trouble with Wilderness." In the course of writing it, I came to see an unconscious beginning, a nudge that set my thoughts in motion. Almost twenty years ago, doing research in the archives of the University of Wisconsin, I ran across a letter Aldo Leopold wrote to William Vogt about the manuscript that became *Road to Survival.* It left out only, Leopold said, the question of "whether the philosophy of industrial culture is not, in its ultimate development, irreconcilable with ecological conservation. I think it is. . . . Industrialism might theoretically be conservative if there were an ethic that limits its application to what does not impair (a) permanence and stability of the land (b) beauty of the land. But there is no such ethic, nor likely to be."[30] Environmentalism, it seems to me, began with that question and aimed to find or make that ethic. In its quest it took, often unconsciously, a religious view, finding the roots of the environmental crisis in our faulty understanding of our species and its place, and asking that we heal ourselves and the land by acknowledging our place and acting in accord with it. Environmentalism has embarked on a great enterprise, the integration into nature (without abandoning human society) of body, mind, spirit, society, and economy. It seeks dreams large enough to inspire individuals and wise enough to guide humanity, dreams that speak to our lives and the wonderful world in which we live them.

NOTES

INTRODUCTION

1. William James, *The Varieties of Religious Experience* (1902; reprint, New York: Random House, 1994), 61, 552. Richard P. McBrien, *Catholicism* (New York: Harper Collins, 1994), 6, spoke of religion at its base, before it achieved organized conviction, as an "understanding, affirmation, and expression about human existence."

2. Thomas Sowell, *Austin-American Statesman,* 27 May 2001. Chuck Cushman of the American Land Rights Association, quoted in *Parade* magazine, *Austin American-Statesman,* 25 November 2001, 7.

3. E. B. White, "A Slight Sound at Evening," in *The Norton Book of Nature Writing,* ed. Robert Finch and John Elder (New York: Norton, 1990), 475–76.

4. Steven Weinberg, *The First Three Minutes* (New York: Basic Books, 1977), 154, 155. On this theme, see also Jacques Monod, *Chance and Necessity* (New York: Vantage, 1974), and Ursula Goodenough, *Sacred Depths of Nature* (New York: Oxford University Press, 1993).

5. Carl Sagan, *Cosmos* (New York: Wing Books, 1980).

6. Michael Ruse, *Can a Darwinian Be a Christian?* (New York: Cambridge University Press, 2001); John Haught, *God After Darwin: A Theology of Evolution.* (Boulder, Colo.: Westview Press, 2000). See also Stephen Jay Gould, *Rocks of Ages* (New York: Ballantine, 1999).

7. Dave Foreman, *Confessions of an Eco-Warrior* (New York: Harmony Books, 1991), 3–4.

8. William James, *Varieties of Religious Experience,* 552.

1 / NEWTON'S DISCIPLES

1. Ted Steinberg, *Down to Earth* (New York: Oxford University Press, 2002), 239. Steinberg includes a good discussion of the context of environmental news in the late 1960s.

2. Blurb on the back of a paperback edition of Rachel Carson, *Silent Spring* (New York: Fawcett, 1970).

3. Rachel Carson, *Silent Spring* (Boston: Houghton Mifflin, 1962), 23.

4. Linda Lear, *Rachel Carson* (New York: Henry Holt, 1997); Thomas R. Dunlap, *DDT* (Princeton, N.J.: Princeton University Press, 1981); Edmund Russell, *War and Nature* (Cambridge, Mass.: Cambridge University Press, 2001).

5. Rachel Carson, *A Sense of Wonder* (New York: Harper Collins, 1999), 9–11.

6. Carson, *A Sense of Wonder,* 54, 106, 101.

7. This last paraphrase is from Annie Dillard, *Pilgrim at Tinker Creek* (New York: Harper and Row, 1974), 8.

8. Carolyn Marvin and David W. Ingle, *Blood Sacrifice and the Nation* (New York: Cambridge University Press, 1999).

9. Peter Gay, *The Enlightenment: An Interpretation: The Science of Freedom* (New York: Knopf, 1969).

10. Alexander Pope, "Epitaph Intended for Sir Isaac Newton," *Bartlett's Familiar Quotations* (New York: Little, Brown, 1980), 340.

11. For an extensive modification, see Stephen Jay Gould, *The Structure of Evolutionary Theory* (Cambridge, Mass.: Harvard University Press, 2002).

12. E. O. Wilson, *On Human Nature* (Cambridge, Mass.: Harvard University Press, 1978), 202, cites God's questions to Job.

13. Andrew Dickson White, *A History of the Warfare of Science with Theology in Christendom* (New York: Appleton, 1896).

14. For a critique of this Darwinian legend, see Stephen Jay Gould, "Knight Takes Bishop?" in *Bully for Brontosaurus* (New York: Norton, 1991), 385–401.

15. There are Catholic priests, Protestant clergy, and Jewish rabbis who are scientists as well. The attraction of this division between science and religion is apparent in its invocation by those who do not believe in the supernatural order. On scientific materialism, see Richard Dawkins, *The Blind Watchmaker* (New York: Norton, 1986); Daniel Dennett, *Darwin's Dangerous Idea* (New York: Simon and Schuster, 1995); E. O. Wilson, *Consilience: The Unity of Knowledge*

(New York, Knopf, 1998); and Stephen Jay Gould, *Rocks of Ages* (New York: Ballantine, 1999). For an overview, see Ian G. Barbour, *Religion and Science: Historical and Contemporary Issues* (San Francisco: Harper Collins, 1997). In *God After Darwin: A Theology of Evolution* (Boulder, Colo.: Westview Press, 2000), John Haught presents a theology incorporating the Darwinian vision.

16. Annie Dillard, *Living by Fiction* (New York: Harper and Row, 1982), 136.

17. Mirceau Eliade, *The Sacred and the Profane* (1957; English edition, New York: Harcourt Brace, 1959), 151.

18. On appreciation as a central activity, see Carl Sagan, *Cosmos* (New York: Random House, 1980), 345. On the critique of Eliade, see Mary Douglas, "Heathen Darkness," in *Implicit Meanings* (London: Routledge and Kegan Paul, 1975), 75–81. Eliade, *The Sacred and the Profane,* 151.

19. E. O. Wilson, *Naturalist* (Washington, D.C.: Island Press, 1994), 43–44. E. O. Wilson, *Biophilia* (Cambridge, Mass.: Harvard University Press, 1984), 1.

20. Wilson, *Biophilia,* 22, 2, 119.

21. E. O. Wilson, *The Future of Life* (New York: Knopf, 2002).

22. Michael Shermer, "The Shamans of Scientism," *Scientific American* 286, no. 6 (June 2002): 35.

23. J. B. S. Haldane, in *Bartlett's Familiar Quotations,* 721.

24. Wilson, *Consilience.* Stephen Jay Gould, *I Have Landed* (New York: Harmony, 2002), 13–14.

25. Michael Ruse, *Can a Darwinian Be a Christian?* (New York: Cambridge University Press, 2001), 186.

26. Sagan, *Cosmos,* 4, 174.

27. Carl Sagan, *The Demon-Haunted World: Science as a Candle in the Dark.* (New York: Random House, 1995), 35, 37–38, 29.

28. Wilson, *Consilience,* 8, 9.

29. Wilson, *Consilience,* 12. A recent counterargument is Wendell Berry, *Life Is a Miracle* (Washington, D.C.: Counterpoint, 2000).

30. Wilson, *Consilience,* 6, 6–7, 7.

31. E. O. Wilson, *Sociobiology* (Cambridge, Mass.: Harvard University Press, 1975); *Biophilia; On Human Nature* (Cambridge, Mass.: Harvard University Press, 1978); *Consilience.*

32. Steven Weinberg, *The First Three Minutes* (New York: Basic Books, 1977),

154, 155. See also Jacques Monod, *Chance and Necessity* (New York: Vantage, 1974).

33. This is a constant theme in the work of Carl Sagan and Stephen Jay Gould, the two most popular science popularizers of the last generation. E. O. Wilson makes the same argument in *Consilience*.

34. Ursula Goodenough, *The Sacred Depths of Nature* (New York: Oxford University Press, 1998), 11, 13.

35. Julian Simon, *The Ultimate Resource* (Princeton: Princeton University Press, 1981).

36. *Updated Last Whole Earth Catalog* (Sausalito, Calif., 1974), also *Essential Whole Earth Catalog* (Garden City, N.Y.: Doubleday, 1986).

37. Wallace Kaufman, *No Turning Back* (New York: Basic Books, 1994), 13, 14.

38. *Economist,* 4 August 2001, 63–65; 2 February 2002, 75. *Wall Street Journal,* 2 October 2001. Bjorn Lomborg, *The Skeptical Environmentalist* (New York: Cambridge University Press, 2001).

39. E. F. Schumacher, *Small Is Beautiful* (New York: Harper, 1973), 39–40.

40. Isaiah Berlin, "The Romantic Revolution," in *The Sense of Reality* (New York: Farrar, Straus and Giroux, 1996), 178.

41. Berlin, "Romantic Revolution," 182–85.

42. This view comes from Ralph Waldo Emerson, "Self-Reliance," in *The Selected Writings of Ralph Waldo Emerson,* ed. Brooks Atkinson (New York: Modern Library, 1992), 134.

43. On this, see Harold Livesay, *American Made* (Boston: Little, Brown, 1979).

44. Berlin, "Romantic Revolution," 178–79.

45. The web sites for these organizations and their publications are worth serious and extended study. On the early movement, see Gerald O'Neill, *The High Frontier* (New York: Morrow, 1977). Later examples include Adrian Berry, *The Giant Leap* (New York: 2001), and the magazine *Artemis.*

46. Sonia Orwell and Ian Angus, eds., *As I Please,* vol. 3 of the *Collected Essays, Journalism and Letters of George Orwell* (New York: Harcourt Brace, 1968), 103; "Looking Back on the Spanish Civil War," in *My Country Right or Left,* volume 2 of *Collected Essays,* 265.

47. Carson, *Silent Spring,* 261.

48. George Perkins Marsh, *Man and Nature* (1864; reprint, Cambridge, Mass.: Harvard University Press, 1965), 36.

49. Liberty Hyde Bailey, *The Holy Earth* (1915; reprint, Ithaca: New York College of Agriculture, 1980), 13.

50. Carolyn Merchant, *The Death of Nature* (San Francisco: Harper and Row, 1980).

51. Peter Hay, *Main Currents in Western Environmental Thought* (Bloomington: Indiana University Press, 2002), 100–106, surveys this dispute.

52. Lynn White, "The Historic Roots of Our Ecologic Crisis," *Science* 155 (10 March 1967): 1203–7.

53. Douglas, "Heathen Darkness," in *Implicit Meanings*, 73.

54. For a recent discussion, incorporating environmental ideas, see Langdon Gilkey, *Nature, Reality, and the Sacred* (Minneapolis: Fortress Books, 1993).

55. For an early critique of White and an approach to the relation between religion and environmental ideas, see David Spring and Eileen Spring, eds., *Ecology and Religion in History* (New York: Harper and Row, 1974). Most commentators on religion and ecological consciousness still deal with White, now usually dismissing his argument as oversimplified.

56. The classic work on natural theology was Gilbert White, *The Natural History of Selbourne* (1787; reprint, New York: Penguin, 1977), still in print. For a secular version, see Stephen Hawking, *The Theory of Everything* (Boston: New Millennium, 2002).

57. Edgar Anderson, *Plants, Man, and Life* (Berkeley: University of California Press, 1971), 45–46.

58. Stephen Jay Gould's long-running series of essays in *Natural History* had as series title Darwin's comment about there being "a certain grandeur in this view of life." Carl Sagan, *Cosmos* and *The Demon-Haunted World*. E. O. Wilson, *The Diversity of Life* (New York: Norton, 1992).

59. John Burroughs, *The Gospel of Nature* (1905; reprint, Bedford, Mass.: Applewood, n.d.); originally published in *Time and Change*. Thomas R. Dunlap, "The Realistic Animal Story: Ernest Thompson Seton, Charles Roberts, and Darwinism," *Forest and Conservation History* 36 (1992): 56–62.

60. John Burroughs, *Light of Day* (Boston: Houghton Mifflin, 1900), 207.

61. The National Wildlife Federation's annual *Conservation Directory* shows national organizations. State and local organizations are harder to trace. Planet Drum and the Turtle Island Bioregional Gatherings, formerly the North American Bioregional Congresses, provide entry into the bioregional debate. *CoEvo-*

lution, a quarterly magazine, was an early focus for published accounts and information about gatherings.

2 / EMERSON'S CHILDREN

1. Aldo Leopold, *A Sand County Almanac* (1949; reprint, New York: Ballantine, 1970), 262.

2. Alfred Russell Wallace, *My Life,* vol. 1 (London: Chapman and Hall, 1905): 192.

3. On Thoreau's nature, see David R. Foster, *Thoreau's Country* (Cambridge, Mass.: Harvard University Press, 1999). On modern naturalists, see, for example, Robert Pyle, *The Thunder Tree: Lessons from an Urban Wildland* (New York: Lyons Press, 1993), and Gary Nabhan, *Cultures of Habitat* (Washington, D.C.: Counterpoint, 1997). Early fascination with nature is a theme in the autobiographies of field naturalists, ecologists, and nature writers.

4. John Muir, *The Mountains of California,* in *John Muir,* ed. William Cronon (New York: Library of America, 1997), 315–547.

5. Catherine Albanese, *Nature Religion in America* (Chicago: University of Chicago Press, 1990), 12.

6. Albanese, *Nature Religion,* 8–11, 201.

7. Albanese, *Nature Religion,* 7–8.

8. Ralph Waldo Emerson, "Nature," in *The Selected Writings of Ralph Waldo Emerson,* ed. Brooks Atkinson (New York: Modern Library, 1992), 6, 15, 17.

9. Mark Stoll, *Protestantism, Capitalism, and Nature in America* (Albuquerque: University of New Mexico Press, 1997).

10. Emerson, *Selected Writings,* 3, 23, 15.

11. Lawrence Buell, *The Environmental Imagination* (Cambridge, Mass.: Harvard University Press, 1995), discusses the rise of the cult. On Thoreau's continuing immediate influence, see Foster, *Thoreau's Country,* written by a man who took to the New England woods in 1977 to build a cabin with his own hands.

12. Henry David Thoreau, *Walden* (1854; reprint, New York: Modern Library, 1937), 81.

13. Eliot Porter, *In Wildness Is the Preservation of the World* (San Francisco: Sierra Club, 1962).

14. For a recent analysis of Thoreau's wilderness and wildness, see Daniel Botkin, *No Man's Garden* (Washington, D.C.: Island Press, 2001).

15. On the need for concrete experience, see Jack Turner, *The Abstract Wild* (Tucson: University of Arizona Press, 1996).

16. Robert D. Richardson, Jr., *Emerson: The Mind on Fire* (Berkeley: University of California Press, 1995), 3.

17. Henry David Thoreau, *The Maine Woods* (New York: Penguin, 1988), 94–95.

18. Suzanne Zeller, *Inventing Canada* (Toronto: University of Toronto Press, 1987), provides a good introduction to natural history's scope and activities.

19. Alexander von Humboldt, *Cosmos: A Sketch of the Physical Description of the Universe* (1845–1862; reprint, Baltimore, Md.: Johns Hopkins University Press, 1997).

20. Laura Dassow Walls, *Seeing New Worlds* (Madison: University of Wisconsin Press, 1995), 134.

21. John Muir, *The Story of My Boyhood and Youth*, in *John Muir: Nature Writings*, ed. William Cronon (Library of America, 1997), 7. On this period, see Steven J. Holmes, *The Young John Muir: An Environmental Biography* (Madison: University of Wisconsin Press, 1999).

22. Muir, *Story of Boyhood and Youth*, 34.

23. Muir, "The Wild Parks and Forest Reservations of the West," in Cronon, *John Muir*, 721.

24. Stoll, *Protestantism, Capitalism, and Nature in America*. Dennis C. Williams, *God's Wilds: John Muir's Vision of Nature* (College Station: Texas A&M University Press, 2002), argues that Muir was a Christian and engages the earlier literature.

25. This is evident in all his extended writings, but an obvious example is the use of biogeographical terms to place the various species of trees in the mountains. See "The Mountains of California," in Cronon, *John Muir*, 403–53.

26. John Muir, "My First Summer in the Sierra," in Cronon, *John Muir*, 245.

27. F. Darwin and A. C. Seward, eds., *More Letters of Charles Darwin*, vol. 1 (London: John Murray, 1903), 94, cited in James R. Moore, "Darwin of Down," *The Darwinian Heritage*, ed. David Kohn (Princeton: Princeton University Press, 1985), 461.

28. Muir, "First Summer," 288.

29. Muir, "First Summer, 263.

30. Muir, "Mountains of California," 482.

31. Muir, "First Summer," 274.

32. Muir, "First Summer," 198.

33. John Burroughs, *The Gospel of Nature* (1905; reprint, Bedford, Mass.: Applewood, n.d.). John Burroughs, *Accepting the Universe* (New York: Russell and Russell, 1920). John Burroughs, *Light of Day* (Boston: Houghton Mifflin, 1900).

34. Burroughs, *Light of Day,* 207.

35. Burroughs, *Gospel of Nature,* 12.

36. Burroughs, *Gospel of Nature,* 32.

37. Burroughs, *Accepting the Universe,* 18.

38. Burroughs, *Gospel of Nature,* 41.

39. Burroughs, "Shall We Accept the Universe?" in *Accepting the Universe,* 5.

40. The literature here is vast. A good introduction to contemporary issues is Ian G. Barbour, *Religion and Science* (New York: Harper Collins, 1997).

41. Ernest Thompson Seton, *The Ten Commandments in the Animal World* (Garden City: Doubleday, 1925). Thomas R. Dunlap, "The Realistic Animal Story: Ernest Thompson Seton, Charles Roberts, and Darwinism," *Forest and Conservation History* 36 (1992): 56–62.

42. For a sample of the Jeffers literature, see James Karman, *Critical Essays on Robinson Jeffers* (Boston: Hall, 1990). On Jeffers's influence, see Karman's introduction to that volume, 1–32, particularly 24–26. A summary and analysis of Jeffers and his influence is in Max Oelschlaeger, *The Idea of Wilderness* (New Haven, Conn.: Yale University Press, 1991), 243–63.

43. Karman, *Critical Essays,* 24–26.

44. David Brower, ed., *Not Man Apart* (San Francisco: Sierra Club, 1965).

45. Karman, *Critical Essays,* 24. William Everson, *The Excesses of God: Robinson Jeffers As a Religious Figure* (Stanford: Stanford University Press, 1988), 82.

46. This argument relies on Everson, *The Excesses of God,* quotes from 2–3, 17, 4.

47. Quoted in Karman, *Critical Essays,* 9.

48. Annie Dillard, *Pilgrim at Tinker Creek* (New York: Harper and Row, 1974), 63.

49. Loren Eiseley, *All the Strange Hours* (New York: Scribner's, 1975), 237–38.

50. Henry Beston, *The Outermost House* (1928; reprint, New York: Ballantine, 1971), ix, 2. Aldo Leopold, "The State of the Profession," *Journal of Wildlife Management* 4 (July 1940): 344.

51. Oelschlaeger, *Idea of Wilderness*, 251.

52. For a recent judgment reinforcing this, see Christopher Cokinos, "Images of Inhumanism," a review of *The Selected Poetry of Robinson Jeffers, Science* 294 (9 November 2001): 1288–89.

53. Brower, *Not Man Apart*, 30.

54. Everson, *The Excesses of God*, 78.

55. George M. Wright, Joseph S. Dixon, and Ben H. Thompson, "Fauna of the National Parks of the United States," National Park Service, Fauna series number 1 (Washington, D.C.: Government Printing Office, 1933), 54.

56. Thomas R. Dunlap, *Saving America's Wildlife* (Princeton: Princeton University Press, 1988), 80–81.

57. George Perkins Marsh, *Man and Nature* (1864; reprint, Cambridge, Mass.: Harvard University Press, 1965), 36, 52.

58. Paul Sears, *Deserts on the March* (Norman: University of Oklahoma Press, 1935), 168, 218.

59. William Vogt, *Road to Survival* (New York: Sloane, 1948), 284, 288.

60. Aldo Leopold, "The Land Ethic," in *Sand County*, 262.

61. Aldo Leopold Papers, University of Wisconsin—Madison Archives, Department of Wildlife Ecology, Diaries and Journals, Misc., 1899–1916, 9/25/10-7, Box 1. Liberty Hyde Bailey, *The Holy Earth* (1915; reprint, Ithaca: New York College of Agriculture, 1980).

62. Bailey, *Holy Earth*, 13, 14, 19.

63. Bailey, *Holy Earth*, 13, 14, 19, 23, 51, 13, 35.

64. Leopold, *Sand County*, 128.

65. Aldo Leopold, "A Criticism of the Booster Spirit," in *The River of the Mother of God and Other Essays*, ed. Susan Flader and J. Baird Callicott (Madison: University of Wisconsin Press, 1991), 98–105.

66. Leopold, *Sand County*, 238.

67. Leopold, *Sand County*, 239, 246, 262.

68. Leopold, *Sand County*, 240, 246.

69. Besides his papers, there is now a book of quotations and commentaries:

Curt Meine and Richard L. Knight, eds., *The Essential Aldo Leopold* (Madison: University of Wisconsin Press, 1999).

70. Leopold, *River of the Mother of God,* 124.

71. Leopold, *Sand County,* 158.

3 / JOURNEY INTO SACRED SPACE

1. Linda Graber, *Wilderness as Sacred Space* (Washington, D.C.: American Association of Geographers, 1976), ix.

2. Robert D. Richardson, Jr., *Emerson: The Mind on Fire* (Berkeley: University of California Press, 1995), 163.

3. J. Baird Callicott and Michael P. Nelson, eds., *The Great New Wilderness Debate* (Athens: University of Georgia Press, 1998).

4. From "The Wilderness and Its Place in Forest Recreational Policy" (1921), in *Aldo Leopold's Southwest,* ed. David E. Brown and Neil B. Carmony (1990; reprint, Albuquerque: University of New Mexico Press, 1995), 148.

5. Aldo Leopold, "The River of the Mother of God," in *The River of the Mother of God and Other Essays,* ed. Susan L. Flader and J. Baird Callicott (Madison: University of Wisconsin Press, 1991), 125.

6. Aldo Leopold, "Conserving the Covered Wagon," in Flader and Callicott, *River of the Mother of God,* 131–32.

7. Michael Cohen, *The History of the Sierra Club* (San Francisco: Sierra Club, 1988), 255.

8. Ronald Tobey, *Saving the Prairies* (Berkeley: University of California Press, 1981). On the influences of the culture on scientists' theories, the best example is Greg Mitman, *State of Nature* (Chicago: University of Chicago Press, 1992).

9. A. G. Tansley, *Practical Plant Ecology: A Guide for Beginners in Field Study of Plant Communities* (London: Allen & Unwin, 1923), 22–23.

10. Sigurd Olson, "Spiritual Aspects of Wilderness," in *The Meaning of Wilderness,* ed. David Backes (Minneapolis: University of Minnesota Press, 2001), 119–20.

11. Sigurd Olson, *The Singing Wilderness* (1956; reprint, New York: Knopf, 1967), 8.

12. Olson, *The Singing Wilderness,* 5–6, 10.

13. Olson, *The Singing Wilderness,* 7.

14. Olson, *The Singing Wilderness*, 130–31.

15. David Backes, *A Wilderness Within* (Minneapolis: University of Minnesota Press, 1997), 59.

16. Backes, *A Wilderness Within*, 61, 281, 290.

17. Backes, *A Wilderness Within*, 186.

18. Backes, *A Wilderness Within*, 250.

19. Aldous Huxley, *The Perennial Philosophy* (1945; reprint, New York: Harper, 1970), vii. On Olson's use, see Backes, *A Wilderness Within*, 287–88.

20. Huxley, *Perennial Philosophy*, 1, 21, 35.

21. Backes, *A Wilderness Within*, 252.

22. Backes, *A Wilderness Within*, 310.

23. Backes, *A Wilderness Within*, 292, 310

24. On the canyon lands as an acquired taste, see Clarence Dutton, *Tertiary History of the Grand Canyon District* (Washington, D.C.: Government Printing Office, 1882).

25. A movement to drain the reservoir still exists. See web site at www.drainit.org.

26. Wallace Stegner, ed., *This is Dinosaur* (New York: Knopf, 1955).

27. Stephanie Mills, *Whatever Happened to Ecology?* (San Francisco: Sierra Club, 1989), 36–37.

28. David Brower, "Introduction," in *The Place No One Knew*, by Eliot Porter (San Francisco: Sierra Club, 1963), 6.

29. Porter, *The Place No One Knew*, 19. On this book, see Cohen, *Sierra Club*, 318–19.

30. Porter, *The Place No One Knew*, 20.

31. Porter, *The Place No One Knew*, 101.

32. Porter, *The Place No One Knew*, 110, 114, 120.

33. On effort and reward, see Sigurd Olson, particularly "Flying In," in *Meaning of Wilderness*, 48–57.

34. Edward Abbey, *Desert Solitaire* (1968: reprint, New York: Ballantine, 1971), xii.

35. John Muir, "My First Summer in the Sierra," in *John Muir*, ed. William Cronon (New York: Library of America, 1997), 245.

36. Abbey, *Desert Solitaire*, 6, 270.

37. Cohen, *Sierra Club*, 347–50.

38. Cohen, *Sierra Club,* 350, 348, 349.

39. Susan Zakin, *Coyotes and Town Dogs* (New York: Penguin, 1993).

40. Zakin, *Coyotes and Town Dogs,* 248, 307, 431, 351.

41. Dave Foreman and Bill Haywood, eds., *Ecodefense: A Field Guide to Monkeywrenching,* 2d ed. (1985; reprint, Tucson, Ariz.: Ned Ludd, 1987).

42. Michael P. Nelson, "An Amalgamation of Wilderness Preservation Arguments," in Callicott and Nelson, *Great Wilderness Debate.*

43. Julia Butterfly Hill, *The Legacy of Luna* (New York: HarperCollins, 2000), 9.

44. Dave Foreman, *Confessions of an Eco-Warrior* (New York: Harmony Books, 1991), 3–4.

45. Backes, "Introduction," in *Meaning of Wilderness,* xxviii.

46. *Environmental History* 1 (January 1996), 29–46; quotes from 37, 30, 32.

47. For a defense of Cronon's position, invoking Muir, see Steven J. Holmes, *The Young John Muir: An Environmental Biography* (Madison: University of Wisconsin Press, 1999) 246–47.

48. Gary Snyder, *Wild Earth* 6 (Winter 1996–97), inside cover, 4.

49. Donald M. Waller, "Wilderness Redux," *Wild Earth* 6 (Winter 1996–97), 38.

50. Gary Snyder, *Wild Earth* 6 (Winter 1996–97), 8–9.

51. Bill Willers, *Wild Earth* 6 (Winter 1996–97), 9, 61.

52. George Sessions, "Reinventing Nature? Losing Wilderness?" *Wild Earth* 6 (winter 1996–97), 52. Ellipsis in original.

53. Callicott, "The Wilderness Idea Revisited," in Callicott and Nelson, *Great Wilderness Debate,* 348.

54. Aldo Leopold, *A Sand Country Almanac* (1949; reprint, New York: Ballantine, 1970), 216, 214–220.

55. Debates may be traced in hiking guides. On recent discussions, see such publications as *High Country News* and *Outside.*

4 / SACRED NATURE ENTERS DAILY LIFE

1. Ted Steinberg, *Down to Earth* (New York: Oxford University Press, 2002), 239–40.

2. Ernest Callenbach, *Ecotopia* (1975; reprint, New York: Bantam Books 1977.)

On the counterculture, see Theodore Roszak, *The Making of a Counterculture* (New York: Doubleday, 1969); Charles A. Reich, *The Greening of America* (New York: Random House, 1970). For a full selection of works in this vein, see successive editions of the *Whole Earth Catalog*.

3. Citations to Paul R. Ehrlich, *The Population Bomb* (New York: Ballantine, 1968); Barry Commoner, *The Closing Circle* (New York: Knopf, 1971); Garrett Hardin, "The Tragedy of the Commons," *Science* 162 (1968): 1243–48; Donella Meadows and others, *The Limits to Growth* (1972; reprint, New York: Universe Books, 1974).

4. A collection of the criticism is H.S.D. Cole, ed., *Models of Doom* (New York: Universe, 1973). Quotations from Donella Meadows, *The Global Citizen* (Washington, D.C.: Island Press, 1991), 11, 32.

5. Bjorn Lomborg, *The Skeptical Environmentalist* (New York: Cambridge University Press, 2001).

6. For information on the numbers of environmental organizations in existence in any given year, see National Wildlife Federation, *Conservation Directory* (Washington, D.C.: National Wildlife Federation, annual).

7. Michael Cohen, *The History of the Sierra Club* (San Francisco: Sierra Club, 1988); Thomas B. Allen, *Guardians of the Wild* (Bloomington: Indiana University Press, 1987); Frank Graham, Jr., *The Audubon Ark* (New York: Knopf, 1990).

8. Mark Sagoff, "Animal Liberation and Environmental Ethics: Bad Marriage, Quick Divorce," *Law Journal* 22, no. 2, 297–307, reprinted in Michael E. Zimmerman and others, eds., *Environmental Philosophy* (Englewood Cliffs, N.J.: Prentice Hall, 1993), 84–94.

9. See, for example, Jennifer Everett, "Environmental Ethics, Animal Welfarism, and the Problem of Predation," *Ethics and the Environment* 6, no. 1 (2001): 42–67. Peter Hay, *Main Currents in Western Environmental Thought* (Bloomington: Indiana University Press, 2002), 62–63.

10. A recent analysis is Andrew Kirk, "Appropriating Technology: The *Whole Earth Catalog* and Counterculture Environmental Politics," *Environmental History* 6 (July 2001): 374–94.

11. Sister Marie Chin, *National Catholic Reporter*, 18 February 2000, 20.

12. *American Heritage Dictionary of the English Language* (New York: American Heritage, 1969); the second definition is a standard theological one. See Richard P. McBrien, *Catholicism* (New York: Harper Collins, 1994), 40.

13. See Michael Brower and Warren Leon, *The Consumer's Guide to Effective Environmental Choices: Practical Advice from the Union of Concerned Scientists* (New York: Three Rivers Press, 1999), which reads like a devotional work addressed to the overzealous.

14. Thomas Haskell, *Objectivity Is Not Neutrality* (Baltimore, Md.: Johns Hopkins University Press, 1998), especially "Capitalism and the Origins of the Humanitarian Sensibility, parts 1 and 2," 235–58, 259–79, suggested this comparison.

15. On formal thought, see Hay, *Main Currents*.

16. Peter Berg, *Reinhabiting a Separate Country* (San Francisco: Planet Drum Foundation, 1978).

17. Jennifer Price, "Looking for Nature at the Mall," in *Uncommon Ground*, ed. William Cronon (New York: Norton, 1995), 186–202.

18. See Jennifer Price, "Looking for Nature at the Mall," and Susan Davis, "Touch the Magic" in *Uncommon Ground*, 186–202, 204–17.

19. See, for instance, Andrew J. Feldman, *The Sierra Club Green Guide* (San Francisco: Sierra Club, 1996), and Brower and Leon, *Consumer's Guide*. This last includes much advice to the scrupulous. Quote from Feldman, *The Sierra Club Green Guide*, 230.

20. This is a trademark of the Fort Howard Corporation.

21. The *Amicus Journal* (now *One-Earth*) is a publication of the Natural Resources Defense Council, New York.

22. Brower and Leon, *Consumer's Guide*, 14, 138, 14, 35.

23. Stephanie Mills, *Whatever Happened to Ecology* (San Francisco: Sierra Club, 1989), and Stephanie Mills, *In Service of the Wild* (Boston: Beacon Press, 1995), deal with these actions.

24. *CoEvolution Quarterly*, no. 32 (1981). *Essential Whole Earth Catalog* (New York: Doubleday, 1986).

25. *Essential Whole Earth Catalog* (1986), 46–49, 43.

26. Mills, *In Service of the Wild*.

27. Gary Snyder, *The Real Work: Interviews and Talks, 1964–1979* (New York: New Directions, 1980), 39, 29–30.

28. Snyder, *The Real Work*, 153, 182, 185.

29. Snyder, *The Real Work*, 82.

30. Snyder, *The Real Work,* 39, 40.

31. Snyder, *The Real Work,* 40, 96.

32. Wendell Berry, "Christianity and the Survival of Creation," in *Sacred Trusts: Essays on Stewardship and Responsibility,* ed. Michael Katakis (San Francisco: Mercury House, 1993), 38, 54; quote on pp. 41–42.

33. Berry, "Christianity and the Survival of Creation," 48–49, 49–50.

34. Wendell Berry, *The Unsettling of America: Culture and Agriculture* (San Francisco: Sierra Club, 1977), 19. A recent statement of this view is his essay "Back to the Land," *Amicus Journal* 20 (Winter 1999): 37–40.

35. Berry, *Unsettling of America,* 41, 43, 48.

36. Aldous Huxley, *The Perennial Philosophy* (1945; reprint, New York: Harper, 1970), 79, 76.

37. Carolyn Merchant, *Radical Ecology* (New York: Routledge, 1992), and Hay, *Main Currents,* provide overviews.

38. Jeffrey C. Ellis, "On the Search for a Root Cause," in Cronon, *Uncommon Ground,* 256–68.

39. Arne Naess, *Ecology, Community, and Lifestyle* (New York: Cambridge University Press, 1989), 1, 9.

40. Bill Devall and George Sessions, *Deep Ecology* (Salt Lake City: Peregrine Smith, 1985), ix–x, 7, 7, 10, 18–24, 81.

41. Devall and Sessions, *Deep Ecology,* 28–29.

42. Bill Devall, *Simple in Means, Rich in Ends* (Salt Lake City: Peregrine Smith, 1988), 25, 35, 201, 12, 12, 13.

43. Aldo Leopold, *A Sand County Almanac* (1949; reprint, New York: Ballantine, 1970), xvii.

44. Author's interview with Joseph J. Hickey, 16 July 1973.

45. David Takacs, *The Idea of Biodiversity* (Baltimore, Md.: Johns Hopkins University Press, 1996). Author's observations.

46. George Orwell, *The Road to Wigan Pier* (1937; reprint, New York: Berkeley, 1961), particularly part 2.

47. William James, *The Varieties of Religious Experience* (1902; reprint, New York: Modern Library, 1994), 569.

48. Reich, *Greening of America,* and Roszak, *The Making of a Counterculture,* were the most visible of this literature.

5 / IN FOR THE LONG HAUL: LIVING IN THE WORLD

1. Mark Dowie, *Losing Ground* (Cambridge, Mass.: MIT University Press, 1995), made one strong argument. Jeffrey C. Ellis, "On the Search for a Root Cause," in *Uncommon Ground,* ed. William Cronon (New York: Norton, 1995), 256–68, analyzed the movement's problems. Carolyn Merchant, *Radical Ecology* (New York: Routledge, 1992), discussed radical positions.

2. Richard P. McBrien, *Catholicism* (New York: Harper Collins, 1994), 883.

3. Donella Meadows and others, *The Limits to Growth* (1972; reprint, New York: Universe, 1974).

4. Meadows and others, *The Limits to Growth,* 21.

5. Donella Meadows, *The Global Citizen* (Washington, D.C.: Island Press, 1991), 6, 11. H.S.D. Cole, ed., *Models of Doom* (New York: Universe, 1973).

6. Meadows, *Global Citizen,* 32.

7. Julian Simon, *The Ultimate Resource* (Princeton: Princeton University Press, 1981).

8. Daniel Botkin, *No Man's Garden* (Washington, D.C.: Island Press, 2001), 50, 51, xiii.

9. A pioneer text on this problem is K. William Kapp, *The Social Costs of Private Enterprise* (Cambrige, Mass.: Harvard University Press, 1950).

10. On the book's popularity, see E. F. Schumacher, *Small Is Beautiful* (Point Roberts, Wash.: Hartley and Marks, 1999), 25th anniversary edition.

11. Schumacher, *Small Is Beautiful,* 27, 41, 277, 277.

12. Schumacher, *Small Is Beautiful,* 49, 52, 58, 105, 277.

13. Schumacher, *Small Is Beautiful,* 281, 280.

14. James Lovelock, *The Ages of Gaia* (New York: Norton, 1988).

15. The literature is still developing. A start on ideas about community is Herman E. Daly and John B. Cobb, Jr., *For the Common Good* (Boston: Beacon, 1994), 397–401. Wendell Berry, *Life Is a Miracle* (Washington, D.C.: Counterpoint, 2000), is a good individual description of the approach to an individual life well lived.

16. Michael Pollan, *Second Nature: A Gardener's Education* (New York: Atlantic Monthly, 1991), 114, 5, 225, 223, 212, 6.

17. Sara Stein, *Noah's Garden: Restoring the Ecology of Our Own Back Yards* (Boston: Houghton Mifflin, 1993).

18. Robert Pyle, *The Thunder Tree: Lessons from an Urban Wildland* (New York: Lyons Press, 1993), xv.

19. Pyle, *Thunder Tree*, 147.

20. Pyle, *Thunder Tree*, xvii, xviii–xix, 152.

21. John W. Brainerd, *The Nature Observer's Handbook* (Chester, Conn.: Globe Pequot Press, 1986), cover blurb.

22. Brainerd, *Nature Observer's Handbook*, 176.

23. Robert Sullivan, *The Meadowlands: Wilderness Adventures at the Edge of a City* (New York: Scribner, 1998), 18, 57, 15, 80, 97.

24. Annie Dillard, *Pilgrim at Tinker Creek* (New York: Harper and Row, 1974), 5–6.

25. The works of Stephen Jay Gould are the best popular modern exposition of that trend. Steven Weinberg, *The First Three Minutes* (New York: Basic Books, 1977). See also Carl Sagan, *The Demon-Haunted World* (New York: Random House, 1995), and E. O. Wilson, *Consilience: The Unity of Knowledge* (New York: Knopf, 1998). Ursula Goodenough, *The Sacred Depths of Nature* (New York: Oxford University Press, 1998). Chet Raymo, *Skeptics and True Believers* (New York: Walker, 1998). An interesting case is Richard Dawkins, *Unweaving the Rainbow* (1998; reprint, Boston: Houghton Mifflin, 2000), an argument for the wonder of nature from a writer committed to complete materialism.

26. E. O. Wilson, *The Future of Life* (New York: Knopf, 2002).

27. William Cronon, ed., *John Muir* (New York: Library of America, 1997).

28. Steven J. Holmes, *The Young John Muir: An Environmental Biography* (Madison: University of Wisconsin Press, 1999), 7, summarizes the range of opinions.

29. On salmon as a symbol, see Joseph E. Taylor III, *Making Salmon* (Seattle: University of Washington Press, 1999), 247–48, more generally, 237–57.

30. For an example of formal theory about humility in environmentalism, see Lisa Gerber, "Standing Humbly before Nature," *Environmental Ethics* 7, no. 1 (2002): 39–53.

31. Derick Jensen, ed., *Listening to the Land* (San Francisco: Sierra Club, 1995).

32. For a version of the bioregional quiz, reprinted from *CoEvolution Quarterly*, see William Cronon, ed., *Uncommon Ground* (New York: Norton, 1995), 369–70. Another quiz appears in the *Essential Whole Earth Catalog* (Garden City, N.Y.: Doubleday, 1986), 46.

33. The quiz does not ask about what earlier generations would have called

woodcraft (a limitation that speaks to environmentalism's origins in the urban middle class). There are no questions on how to build a fire, make a camp, or even catch a fish. Nor does it consider knowledge about making a living in nature— no questions about planting dates for crops, what trees in the area make good lumber or firewood, or what furbearers live in the streams and woods.

34. Rachel Carson, *Silent Spring* (Boston: Houghton Mifflin, 1962), 261.

35. Gregg Easterbrook, *A Moment on the Earth* (New York: Viking, 1995), xv, 649, 670–71. Paul R. Ehrlich and Anne H. Ehrlich, *Betrayal of Science and Reason* (1996; reprint, Washington, D.C.: Island Press, 1998), give a good sample of the environmental attack.

36. Bjorn Lomborg, *The Skeptical Environmentalist* (New York: Cambridge University Press, 2001). A Web search will give a large sampling of opinions.

37. McBrien, *Catholicism*, 935. Gary Snyder, "The Real Work," in *The Real Work: Interviews and Talks, 1964–1979*, ed. William Scott McLean (New York: New Directions, 1980), 82.

38. Bill McKibben, *Hope, Human and Wild* (Boston: Little, Brown, 1995).

39. McKibben, *Hope, Human and Wild*, 3. Bill McKibben, *The End of Nature* (New York: Random House, 1989).

40. McKibben, *Hope, Human and Wild*, 11, 3, 35, 36, 127, 193.

41. McKibben, *Hope, Human and Wild*, 184, 3, 204. Italics in original.

42. McKibben, *Hope, Human and Wild*, 203, 208, 214, 226.

CONCLUSION "QUO VADIS?"

1. Jennifer Price, *Flight Maps* (New York: Basic Books, 1999); Susan Davis, *Spectacular Nature* (Berkeley: University of California Press, 1997). The newspaper ad appeared in the Bryan, Texas, *Eagle*.

2. *National Wildlife Catalog*, spring 1999.

3. David Takacs, *The Idea of Biodiversity* (Baltimore, Md.: Johns Hopkins University Press, 1996).

4. On the meanings of freedom, see Eric Foner, *The Story of American Freedom* (New York: Norton, 1998).

5. On action before understanding, see William James, *The Will to Believe and Other Essays in Popular Philosophy* (1897; reprint, New York: Dover, 1956), and comments on that in Charles Taylor, *Varieties of Religion Today: William James*

Revisited (Cambridge, Mass.: Harvard University Press, 2002), 45–51. On the theological idea of praxis, see Richard P. McBrien, *Catholicism* (New York: Harper Collins, 1994), 883.

6. Peter Berg, *Reinhabiting a Separate Country* (San Francisco: Planet Drum Foundation, 1978), 1.

7. Wendell Berry, *Jayber Crow* (Washington, D.C.: Counterpoint, 2000), 254.

8. Henry David Thoreau, "Walking," in *Walden and Other Writings of Henry David Thoreau* (New York: Random House, 1937), 597, 597–632.

9. On this search in the environmental religion, see Gary Snyder, *The Practice of the Wild* (San Francisco: North Point Press, 1990).

10. Daniel Botkin, *No Man's Garden* (Washington, D.C.: Island Press, 2001).

11. Botkin, *No Man's Garden*, 171.

12. Botkin, *No Man's Garden*, xiii, xv, xvi, xviii, xx.

13. Botkin, *No Man's Garden*, 51.

14. For a summary, see Carolyn Merchant, *Radical Ecology* (New York: Routledge, 1992).

15. Richard White, "Are You an Environmentalist or Do You Work for a Living?" and James D. Proctor, "Whose Nature? The Contested Moral Terrain of Ancient Forests," both in *Uncommon Ground,* ed., William Cronon (New York: Norton, 1995), 171–85, 269–97.

16. See Dan Simmons, *Hyperion* (New York: Bantam, 1990).

17. See, for example, Dieter T. Hessel and Rosemary Radford Reuther, eds., *Christianity and Ecology* (Cambridge, Mass.: Harvard University Press and Harvard Center for the Study of World Religions, 2000), and Harold Coward and Daniel C. Maguire, eds., *Visions of a New Earth: Religious Perspectives on Population, Consumption, and Ecology* (Albany: State University of New York Press, 2000).

18. On the general shift in religious denominations, see a special issue of *Daedalus,* "Religion and Ecology: Can the Climate Change?" 130, no. 4 (Fall 2001), or John E. Carroll, Paul Brockelman, and Mary Westfall, eds., *The Greening of Faith* (Hanover, N.H.: University Press of New England, 2001).

19. Matt Biers-Ariel, Deborah Newbrun, and Michael Fox Smart, *Spirit in Nature: Teaching Judaism and Ecology on the Trail* (n.p.: Behrman House, 2000).

20. Biers-Ariel, Newbrun, and Smart, *Judaism and Ecology on the Trail,* ix.

21. Biers-Ariel, Newbrun, and Smart, *Judaism and Ecology on the Trail,* 62–64.

22. Shannon Jung, *We Are Home: A Spirituality of the Environment* (Mahwah, N.J.: Paulist Press, 1993).

23. Freeman Dyson, *Disturbing the Universe* (San Francisco: Harpers, 1969), 245. Mary Evelyn Tucker and John A. Grim, "Introduction: The Emerging Alliance of World Religions and Ecology," in *Daedalus* 130 (2001): 9. The entire issue is of interest. See also Stephen R. Kellert and Timothy J. Farnham, eds., *The Good in Nature and Humanity: Connecting Science, Religion, and Spirituality* (Washington, D.C.: Island Press, 2002).

24. Mark Stoll, *Protestantism, Capitalism, and Nature in America* (Albuquerque: University of New Mexico Press, 1997). A closer look at Protestant influence is Stephen J. Holmes, *The Young John Muir: An Environmental Biography* (Madison: University of Wisconsin Press, 1999).

25. Mark Stoll, "Green versus Green: Religions, Ethics, and the Bookchin-Foreman Debate," *Environmental History* 6 (July 2001), 423.

26. Barry Commoner, *The Closing Circle* (New York: Knopf, 1971), 37. James Lovelock, *The Ages of Gaia* (New York: Norton, 1988), and *Healing Gaia* (New York: Harmony, 1991).

27. See, for example, Donald Worster, "The Ecology of Order and Chaos," in Donald Worster, *The Wealth of Nature* (New York: Oxford University Press, 1993), 156–70.

28. An example of this approach is John Haught, *God After Darwin* (Boulder, Colo.: Westview Press, 2000).

29. Kathleen Norris, *The Cloister Walk* (New York: Riverhead, 1996), 94.

30. Aldo Leopold to William Vogt, Aldo Leopold Papers, University of Wisconsin—Madison Archives, Department of Wildlife Ecology, 10-2, 4.

SELECTED BIBLIOGRAPHY

Abbey, Edward. *Desert Solitaire.* 1968. Reprint, New York: Ballantine, 1971.

———. *The Monkey Wrench Gang.*

Albanese, Catherine. *Nature Religion in America.* Chicago: University of Chicago Press, 1990.

Backes, David. *A Wilderness Within.* Minneapolis: University of Minnesota Press, 1997.

Bailey, Liberty Hyde. *The Holy Earth.* 1915. Reprint, Ithaca, N.Y: New York College of Agriculture, 1980.

Barbour, Ian G. *Religion and Science: Historical and Contemporary Issues.* New York: HarperCollins, 1997.

Berg, Peter. *Reinhabiting a Separate Country.* San Francisco: Planet Drum Foundation, 1978.

Berry, Wendell. *Jayber Crow.* Washington, D.C.: Counterpoint, 2000.

———. *Life Is a Miracle.* Washington, D.C.: Counterpoint, 2000.

———. *Meeting the Expectations of the Land.* San Francisco: North Point Press, 1984.

———. *The Unsettling of America: Culture and Agriculture.* San Francisco: Sierra Club, 1977.

Beston, Henry. *The Outermost House,* 1929. Reprint, New York: Ballantine, 1971.

Biers-Ariel, Matt, Deborah Newbrun, and Michael Fox Smart. *Spirit in Nature: Teaching Judaism and Ecology on the Trail.* N.p.: Behrman House, 2000.

Botkin, Daniel. *No Man's Garden.* Washington, D.C.: Island Press, 2001.

Brainerd, John W. *The Nature Observer's Handbook.* Chester, Conn.: Globe Pequot Press, 1986.

Bibliography

Brower, David, ed. *Not Man Apart.* San Francisco: Sierra Club, 1965.

Brower, Michael, and Warren Leon. *The Consumer's Guide to Effective Environmental Choices: Practical Advice from the Union of Concerned Scientists.* New York: Three Rivers Press, 1999.

Burroughs, John. *Accepting the Universe.* New York: Russell and Russell, 1920.

————. *The Gospel of Nature.* 1905. Reprint, Bedford, Mass.: Applewood, n.d.

————. *Light of Day.* Boston: Houghton Mifflin, 1900.

Callenbach, Ernest. *Ecotopia.* 1975. Reprint, New York: Bantam Books, 1977.

Callicott, J. Baird. *In Defense of the Land Ethic.* Albany: State University of New York Press, 1989.

————. *Beyond the Land Ethic.* Albany: State University of New York Press, 1999.

Callicott, J. Baird, and Michael P. Nelson, eds. *The Great New Wilderness Debate.* Athens: University of Georgia Press, 1998.

Carroll, John E., Paul Brockelman, and Mary Westfall, eds. *The Greening of Faith.* Hanover, N.H.: University Press of New England, 2001.

Carson, Rachel. *A Sense of Wonder.* New York: Harper Collins, 1999.

————. *Silent Spring.* 1962. Reprint, New York: Fawcett, 1970.

Commoner, Barry. *The Closing Circle.* New York: Knopf, 1971.

Coward, Harold, and Daniel C. Maguire, eds. *Visions of a New Earth: Religious Perspectives on Population, Consumption, and Ecology.* Albany: State University of New York Press, 2000.

Cronon, William, ed. *Uncommon Ground.* New York: Norton, 1995.

Daedalus. Special issue, "Religion and Ecology: Can the Climate Change?" 130, no. 4 (Fall 2001).

Daly, Herman E., and John B. Cobb, Jr. *For the Common Good.* Boston: Beacon, 1994.

Dawkins, Richard. *The Blind Watchmaker.* New York: Norton, 1986.

————. *Unweaving the Rainbow.* Boston: Houghton Mifflin, 1998.

Dennett, Daniel. *Darwin's Dangerous Idea.* New York: Simon and Schuster, 1995.

Devall, Bill. *Simple in Means, Rich in Ends.* Salt Lake City: Peregrine Smith, 1988.

Devall, Bill, and George Sessions. *Deep Ecology.* Salt Lake City: Peregrine Smith, 1985.

Dillard, Annie. *Pilgrim at Tinker Creek.* New York: Harper and Row, 1974.

Easterbrook, Gregg. *A Moment on the Earth.* New York: Viking, 1995.

Ehrlich, Paul. *The Population Bomb.* New York: Ballantine, 1968.

Bibliography

Eliade, Mirceau. *The Sacred and the Profane.* 1957. English edition. New York: Harcourt Brace, 1959.

Everett, Jennifer. "Environmental Ethics, Animal Welfarism, and the Problem of Predation." *Ethics and the Environment* 6, no. 1 (2000): 42–67.

Everson, William. *The Excesses of God: Robinson Jeffers as a Religious Figure.* Stanford: Stanford University Press, 1988.

Feldman, Andrew J. *The Sierra Club Green Guide.* San Francisco: Sierra Club, 1996.

Foreman, Dave. *Confessions of an Eco-Warrior.* New York: Harmony Books, 1991.

Foreman, Dave, and Bill Haywood. *Ecodefense: A Field Guide to Monkeywrenching.* 2d. ed. 1985. Tucson, Ariz.: Ned Ludd, 1987.

Gerber, Lisa. "Standing Humbly before Nature." *Environmental Ethics* 7, no. 1 (2002): 39–53.

Gilkey, Langdon. *Nature, Reality, and the Sacred.* Minneapolis: Fortress Books, 1993.

Goodenough, Ursula. *The Sacred Depths of Nature.* New York: Oxford University Press, 1998.

Gould, Stephen Jay. *Rocks of Ages.* New York: Ballantine, 1999.

Hardin, Garrett. "The Tragedy of the Commons." *Science* 162 (1968): 1243–48.

Haught, John. *God After Darwin: A Theology of Evolution.* Boulder, Colo.: Westview Press, 2000.

Hessel, Dieter T., and Rosemary Radford Reuther, eds. *Christianity and Ecology.* Cambridge, Mass.: Harvard University Press and Harvard Center for the Study of World Religions, 2000.

Holmes, Steven J. *The Young John Muir: An Environmental Biography.* Madison: University of Wisconsin Press, 1999.

Humboldt, Alexander von. *Cosmos.* 1845–1862. Reprint, Baltimore: Johns Hopkins University Press, 1997.

Huxley, Aldous. *The Perennial Philosophy.* 1945. Reprint, New York: Harper, 1970.

James, William. *The Varieties of Religious Experience.* 1902. Reprint, New York: Modern Library, 1994.

———. *The Will to Believe and Other Essays in Popular Philosophy.* 1897. Reprint, New York: Dover, 1956.

Jensen, Derick, ed. *Listening to the Land.* San Francisco: Sierra Club, 1995.

Jung, Shannon. *We Are Home: A Spirituality of the Environment.* Mahwah, N.J.: Paulist Press, 1993.

Bibliography

Katakis, Michael, ed. *Sacred Trusts: Essays on Stewardship and Responsibility.* San Francisco: Mercury House, 1993.

Kaufman, Wallace. *No Turning Back.* New York: Basic Books, 1994.

Kellert, Stephen R., and Timothy J. Farnham, eds. *The Good in Nature and Humanity: Connecting Science, Religion, and Spirituality.* Washington, D.C.: Island Press, 2002.

Knight, Richard L., and Suzanne Riedel, eds. *Aldo Leopold and the Ecological Conscience.* New York: Oxford University Press, 2002.

Leopold, Aldo. *A Sand County Almanac.* 1949. Reprint, New York: Ballantine, 1970.

Lomborg, Bjorn. *The Skeptical Environmentalist.* New York: Cambridge University Press, 2001.

Lovelock, James. *The Ages of Gaia.* New York: Norton, 1988.

Marsh, George Perkins. *Man and Nature.* 1864. Reprint, Cambridge, Mass.: Harvard University Press, 1965.

McKibben, Bill. *The End of Nature.* New York: Random House, 1989.

———. *Hope, Human and Wild.* Boston: Little, Brown, 1995.

Meadows, Donella, and others. *The Limits to Growth.* 1972. Reprint, New York: Universe Books, 1974.

Meine, Curt, and Richard L. Knight, eds. *The Essential Aldo Leopold.* Madison: University of Wisconsin Press, 1999.

Merchant, Carolyn. *The Death of Nature.* San Francisco: Harper and Row, 1980.

Mitman, Greg. *State of Nature.* Chicago: University of Chicago Press, 1992.

Monod, Jacques. *Chance and Necessity.* New York: Vantage, 1974.

Muir, John. *John Muir.* New York: Library of America, 1997.

Naess, Arne. *Ecology, Community, and Lifestyle.* New York: Cambridge University Press, 1989.

Olson, Sigurd. *The Meaning of Wilderness.* Minneapolis: University of Minnesota Press, 2001.

———. *The Singing Wilderness.* Edited by David Backes. 1956. Reprint, New York: Knopf, 1967.

Philosophy in the Contemporary World 8, no. 2 (Fall/Winter 2001).

Pollan, Michael. *Second Nature: A Gardener's Education.* New York, Atlantic Monthly, 1991.

Porter, Eliot. *In Wilderness Is the Preservation of the World.* San Francisco: Sierra Club, 1962.

————. *The Place No One Knew.* San Francisco: Sierra Club, 1963.

Pyle, Robert. *The Thunder Tree: Lessons from an Urban Wildland.* New York: Lyons Press, 1993.

Raymo, Chet. *Skeptics and True Believers.* New York: Walker, 1998.

Ruse, Michael. *Can a Darwinian Be a Christian?* New York: Cambridge University Press, 2001.

Sagan, Carl. *Cosmos.* New York: Wing Books, 1980.

————. *The Demon-Haunted World: Science as a Candle in the Dark.* New York: Random House, 1995.

Sagoff, Mark. "Animal Liberation and Environmental Ethics: Bad Marriage, Quick Divorce." *Law Journal* 22, no. 2: 297–307. Reprinted in Michael E. Zimmerman and others, eds. *Environmental Philosophy.* Englewood Cliffs, N.J.: Prentice Hall, 1993.

Schumacher, E. F. *Small Is Beautiful.* New York: Harper, 1973.

Sears, Paul. *Deserts on the March.* Norman: University of Oklahoma Press, 1935.

Simon, Julian. *The Ultimate Resource.* Princeton: Princeton University Press, 1981.

Snyder, Gary. *The Real Work: Interviews and Talks, 1964–1979.* New York: New Directions, 1980.

————. *The Practice of the Wild.* San Francisco: North Point Press, 1990.

Stegner, Wallace, ed. *This Is Dinosaur.* New York: Knopf, 1955.

Stein, Sara. *Noah's Garden: Restoring the Ecology of Our Own Back Yards.* Boston: Houghton Mifflin, 1993.

Stoll, Mark. "Green versus Green: Religions, Ethics, and the Bookchin-Foreman Debate." *Environmental History* 6 (July 2001): 412–27.

————. *Protestantism, Capitalism, and Nature in America.* Albuquerque: University of New Mexico Press, 1997.

Sullivan, Robert. *The Meadowlands: Wilderness Adventures at the Edge of a City.* New York: Scribner, 1998.

Turner, Jack. *The Abstract Wild.* Tucson: University of Arizona Press, 1996.

Vogt, William. *Road to Survival.* New York: Sloane, 1948.

Walls, Laura Dassow. *Seeing New Worlds: Henry David Thoreau and Nineteenth-Century Natural Science.* Madison: University of Wisconsin Press, 1995.

Weinberg, Steven. *The First Three Minutes.* New York: Basic Books, 1977.

White, Lynn. "The Historic Roots of Our Ecologic Crisis." *Science* 155 (10 March 1967): 1203–7.

Bibliography

Wild Earth 6 (Winter 1996–1997).

Williams, Dennis C. *God's Wilds: John Muir's Vision of Nature*. College Station: Texas A&M University Press, 2002.

Wilson, E. O. *Biophilia*. Cambridge, Mass.: Harvard University Press, 1984.

———. *Consilience: The Unity of Knowledge*. New York: Knopf, 1998.

———. *The Diversity of Life*. New York: Norton, 1992.

———. *The Future of Life*. New York: Knopf, 2002.

Worster, Donald. *The Wealth of Nature*. New York: Oxford University Press, 1993.

Wu, J., and O. J. Loucks. "From Balance of Nature to Hierarchical Patch Dynamics: A Paradigm Shift in Ecology." *Quarterly Review of Biology* 70 (1995): 439–66.

INDEX

Index